The Training of
CANCER RESEARCHERS

The Training of
CANCER RESEARCHERS

Jose Russo

Fox Chase Cancer Center–Temple Health
Philadelphia, USA

NEW JERSEY · LONDON · SINGAPORE · BEIJING · SHANGHAI · HONG KONG · TAIPEI · CHENNAI · TOKYO

Published by

World Scientific Publishing Co. Pte. Ltd.

5 Toh Tuck Link, Singapore 596224

USA office: 27 Warren Street, Suite 401-402, Hackensack, NJ 07601

UK office: 57 Shelton Street, Covent Garden, London WC2H 9HE

Library of Congress Cataloging-in-Publication Data

Names: Russo, Jose, 1942–   author.
Title: The training of cancer researchers / Jose Russo, Fox Chase Cancer Center, USA.
Description: New Jersey : World Scientific, 2017. | Includes bibliographical
      references and index.
Identifiers: LCCN 2017003571 | ISBN 9789813203143 (hardcover : alk. paper)
Subjects: LCSH: Cancer--Research. | Vocational guidance.
Classification: LCC RC267 .R87 2017 | DDC 616.99/40072--dc23
LC record available at https://lccn.loc.gov/2017003571

British Library Cataloguing-in-Publication Data
A catalogue record for this book is available from the British Library.

Typeset by Stallion Press
Email: enquiries@stallionpress.com

Printed in Singapore

To my daughter Patricia

Preface

This book explores the human origins of the scientific endeavor as related to cancer research from a cultural and historical perspective. In our life time we have seen science grow and expand and cancer research has had to answer increasingly difficult questions with methods that were unimaginable just a few years ago. One important aspect of the cancer research endeavor is the training of the future cancer researchers considering the framing of social, economic, and cultural changes of the human population. Despite the brilliant technology and laser-pointed specializations, still researchers are urged to think in translation faster than ever before turning the new discoveries to the practical situation of each individual cancer. In this book, we will discuss a number of topics that over the years have become increasingly relevant for the training of future cancer researchers to the entire scientific enterprise.

Prof. Jose Russo, MD, FACP

Acknowledgments

My specific acknowledgement and thanks to Patricia A. Russo for her insightful editorial suggestions, critiques, and the delightful moments spent with her discussing the manuscripts and its ideas.

A special thanks to Linda Cathay for helping me in the formatting and downloading of all the manuscripts.

My thanks also to Pathology Consultation Service in Rydal, PA, that financed the writing and editing of this book.

Prof. Jose Russo, MD, FACP

About the Author

Professor Jose Russo, MD, is a Senior Member and Director of the Irma H Russo, MD, Breast Cancer Research Laboratory, Director of the Breast Cancer and The Environment Research Center at the Fox Chase Cancer Center and Adjunct Professor of Pathology and Cell Biology at Jefferson Medical School, and Professor of Biochemistry in Temple Medical School in Philadelphia, Pennsylvania. He has authored more than 400 publications; 13 books, and is a member of several editorial boards of scientific journals. He has received numerous research awards from the National Cancer Institute of the National Institute of Health (NIH) of the United States, from the American Cancer Society and the Department of Defense for his original research on breast cancer. For the last 40 years, he has been an active member of the NIH peer review system and has served as a special reviewer for the American Cancer Society, National Science Foundation, Department of Defense and Veteran Affairs. He has trained 55 PhD and MD investigators in cancer research. The interest of Dr. Russo has a broad base, but with a focused goal (1) to understand the mechanisms that control the susceptibility of the breast epithelium to undergo neoplastic transformation, (2) to identify markers of susceptibility, and (3) to develop strategies for breast cancer prevention.

Contents

Chapter 1

The Original Idea as the Main Driver in Cancer Research

1.1. The Importance of the Individual Mind

In 1946, Dr. Stanley P. Reimann, director of The Research Institute in Philadelphia (presently the Fox Chase Cancer Center–Temple Health), delivered a speech at the third annual meeting of the American-Soviet Medical Society in New York City. He stated, "Some people say that the day of the lone experimenter with a rabbit in one hand and a hypodermic syringe in the other is over. It may well be that the field is still open for the lone experimenter to make discoveries". There is no doubt that both views still have supporters. However, when we study the trend for human endeavors in the sciences, and particularly in cancer research, we observe fewer single-investigator studies and more often the teamwork of a few collaborators — e.g., a paper published in the journal *Genes Genomes Genetics* names 1014 authors with more than 900 undergraduate students among them. Although some questioned whether every person made enough of a contribution to be credited as an author, the paper's senior author, Dr. Sarah Elgin, at Washington University in St. Louis, Missouri, said, "large collaborations with correspondingly large author lists have become a fact of life in genomic research. Putting together the efforts of many people allows you to do good projects".

Yet how will the new generation of leading cancer researchers emerge in this environment where the research endeavor has progressed from the individual effort to the institutional product? And how will national groups address the need for international collaborations between different research teams? How true is Reimann's observation today? A study published few years ago (Adams, 2011) showed that "about 75% of the research output of China, Brazil, India and South Korea remains entirely domestic. The total volume of papers from these four countries has increased 20-fold — from fewer than 15,000 papers annually in 1981 to more than 300,000 papers in 2011. For established economies after the mid-1990s, the domestic research output of the United Kingdom (47,500 papers per year), Germany (45,000 papers) and France (30,000 papers) levelled off while international collaboration in these countries increased more than ten-fold. The author of this article (Adams, 2011) concluded that exceptional research groups share ideas, resources and outcomes. For example, the most frequent international partners of the University of Cambridge, UK, are the Max Planck institutes in Germany, the Massachusetts Institute of Technology and Harvard University, both in Cambridge; the California Institute of Technology in Pasadena; the University of California, Berkeley; and the universities of Toronto, Heidelberg and Tokyo. Harvard's frequent international partners are Imperial College London, University College London, the Max Planck institutes, the Karolinska Institute in Stockholm and the universities of Cambridge, Toronto and Geneva. The conclusion is that internationally co-authored papers are more highly cited because the authors are more likely to be doing excellent research". The counterpart of this is that many outstanding scientists in developing countries will be left out.

All these data can be demoralizing for those who aspire to be cancer researchers and occupy the mind with questions of individual versus teamwork versus large collaborations and whether or not to be part of a larger whole as opposed to being the driver in the research process. Therefore, the main question is: what has been the driving force in cancer research (as well as for all the sciences) to get us where we are today? *And the simple answer is that the research idea germinates in the mind of a single individual most of the time and it is the idea that counts.* At the end, the great triumphs of science have resulted from the determination to follow an individual's idea, as with Marie Curie's discovery of radium or Isaac Newton's discovery of gravitational law (Russo, 2010). It required the industrious and clear mind of Rudolf Virchow to start a systematic review of pathological lesions as demonstrated in the *Handbook of Special Pathology and Therapeutics* in 1847. In this book, Virchow clarified his views on cellular pathology at a critical time. In the early period of pathology, many disease processes were still poorly understood. Many prominent pathologists believed that pathological changes occurred due to an imbalance in the blood of substances, such as fibrin and albumin, which created a "blastema" that formed abnormal cells leading to disease. Virchow, together with Robert Remak (Titford, 2010), categorically stated that cells were derived from other cells, and therefore, pathological cells were also derived from other pathological cells. This innovative concept was later expanded in his *Cellular Pathology*, followed by a three-volume series on tumors in 1863.

Virchow's greatest achievements were in microscopic pathology. Virchow was not the first to study diseased tissues microscopically, but he was the first to recommend a systemic microscopic study of tissues. Likewise, Virchow recommended a complete autopsy; in contrast, previous pathologists only studied specific tissues and organs as directed by clinicians. Virchow established the relationship between cancer and inflammation. He also suggested that chronic inflammation influenced tumor development. Balkwill and Mantovani metaphorically summarized Virchow's idea: "If genetic damage is the match that lights the fire of cancer, inflammation provides the fuel that feeds the flames" (Balkwill and Mantovani, 2001). The importance of inflammation and immunosurveillance for tumor progression is one of the current accepted concepts in cancer (Hanahan and Weinberg, 2011).

In 1914, Theodor Boveri, a disciple of Virchow's, pointed out the role of the chromosomes' constitution of the cell, specifically noting that kariological disorder is initiated by abnormalities of mitosis and that centriolar malfunction might sometimes be involved. Yet his work has only recently been rescued by our increased understanding of the role of genes in cancer (Harris, 2008). Boveri postulated that malignant tumors are clonal outgrowths and that their ability to multiply exponentially is the hallmark of the neoplastic process. These pivotal ideas developed by a single individual are the cornerstone of the genetic theory of cancer.

Another individual who changed our way of thinking in the neurosciences was Santiago Ramón y Cajal, whose ideas pioneered investigations of the microscopic structure of the brain. His unique drawings illustrating the delicate arborizations of brain cells — which he painstakingly extracted from his own histological preparations — allowed him to provide an understanding of the nervous system that is still a source of inspiration for the new cadre of researchers in that field. He discovered the axonal growth cone and demonstrated experimentally that the relationship between nerve cells was not continuous, but contiguous. This provided definitive evidence for what would later be known as the "neuron doctrine", which displaced the concept of reticular theory that was widely accepted in his time.

It is also true that new ideas are often inspired by the environment that the scientist works in, such as a mentor or a reading that stimulates the individual to think in a different direction. For example, Levi-Montalcini read an article by Viktor Hamburger that indicated that when the growing limbs of chick embryos were cut off, the resultant atrophy in the neuronal cell clusters intended to innervate them was due to the loss of an "inductive factor" from the absent limbs. Hamburger suggested that this factor was necessary for the growth and differentiation of the neural precursor cells. Levi-Montalcini repeated this experiment and concluded instead that the neuronal death resulted from the absence of a growth-promoting substance that resulted to be the neural growth factor (Bradshaw, 2013). Again in this example, it is initiative and individual interpretation that make the final mark.

The significant collaboration of more than one mind is also well documented, like in Nirenberg and Matthaei's discovery that RNA, rather than DNA, programmed the synthesis of proteins (Caskey, 2011). These initial discoveries paved the way for finally deciphering the genetic code, making possible the spectacular era of DNA sequencing, recombinant DNA technology and genome projects that followed.

In the same line of thinking, Siekevitz and Zamecnik obtained the definitive data on protein biosynthesis through the use of a cell-free (*in vitro*) system, thus revealing the enzymatic activation of amino acids, the ribosome as the site of peptide-bond formation and the existence of transfer RNA (Pederson and Paul, 2009). In other collaborative situations, one is the main investigator and one makes the crucial observations. This is the case of Elizabeth Carswell and Barbara Williamson who were working in the laboratory of Lloyd J. Old. They noted that tumors of mice turned black, which eventually led to the identification of the cytokine, tumor necrosis factor (which is involved in tumor regression) (Sharma and Allison, 2012).

Another good demonstration of **the idea** of an individual behind important discoveries is the work of Philip Lawley, which laid the foundation that cancer is a genetic disease. He provided the first convincing evidence that DNA is the key target for chemicals that cause cancer, and identified a major DNA-repair mechanism that counteracts the assault of carcinogens on DNA (Venitt and Phillips, 2012). He studied the physico-chemical properties of DNA (before its double-helical structure was revealed by James Watson and Francis Crick). Lawley demonstrated that alkylating agents could bond covalently with DNA to produce stable adducts, a radical idea at a time when it was believed that such interactions were weak or reversible and that proteins were the crucial target of carcinogens. Lawley went on to show how point mutations are induced when potent alkylating mutagens, such as *N*-methyl-*N*-nitrosourea, react with those atoms in DNA that determine base pairing during DNA replication. Point mutations are now known to occur frequently in a variety of human cancer genes.

I cannot emphasize enough the importance of the individual idea and work, which have generated pivotal changes in our understanding of cancer. I could continue enumerating hundreds of cases that support this point but that would transform this chapter into a biographical compendium.

1.2. The Challenges Ahead for Developing the Driving Idea

Next I would like to examine the challenges that the new cadre of cancer researchers will need to face in order to develop the driving idea for solving the problems of cancer. What we know thus far is that by the time most cancers are detected, at least one tumor has grown to contain a billion cells which, by mutation and natural selection, have become altered in ways that allow them to escape the body's safeguard mechanisms. Cancer cells are constantly changing and altering the genes controlling the normal process of growing. In 1900, the leading causes of death in the United States were pneumonia, influenza, and tuberculosis; a century later, they are heart disease and cancer. More than 40 years since "war" was declared on cancer, 15% of deaths worldwide are attributable to cancer. In *The Emperor of All Maladies*, a beautiful book written by Mukherjee (2010), there is a thorough description of the history of cancer. In the nineteenth and early twentieth centuries, cancer was defined by the demonstration of invasion and metastases, based on gross findings at surgery or autopsy. Although histopathologic examination of tumors — focusing on tissue changes — became possible with greater and greater resolution over time, the definition of cancer remained the same and prognostication based on histopathologic analyses of tumor biopsies and resection specimens was still not possible. When the concepts of tumor grading and staging were discovered in the 1920s and 1930s, histopathology could finally provide prognostic information, fostering the tumor staging and eventually the cancer biomarker fields. Analyzing such factors as the tumor histologic pattern, degree of nuclear pleomorphism, number of mitoses, presence of inflammatory response, blood vessel invasion, lymph node involvement, and the presence of hormone receptors (Bloom and Richardson, 1957; Bloom *et al.*, 1970; Russo *et al.*, 1987), attempts have been made to identify patients at high risk of early disease recurrence and thus more effectively target aggressive auxiliary chemotherapy and intensive follow-up protocols. However, many cancers having similar histologic patterns differ markedly in their recurrence behavior, which affects a patient's survival. In the case of breast cancer it was obvious that single morphologic or biologic characteristics were insufficient to predict the biologic behavior of a tumor; therefore, a combination of various criteria was necessary to accurately identify subpopulations of patients, such as breast cancer sufferers at increased risk of recurrence or shortened survival. For example, after the publication of the new genomic classification of breast cancer in 2006 (Sørlie *et al.*, 2006), several adaptations, compiling a variety of representative genes, have been introduced in the practice of oncology. Among these adaptations is the Oncotype Dx, which contains a 21-gene expression signature (Mamounas *et al.*, 2010). This study comprises female breast cancer patients that are estrogen receptor positive (ER+) and lymph node negative (LN–). The Oncotype was developed from a formalin-fixed paraffin-embedded tissue (FFPE) assay to predict distant recurrence of ER+ breast cancer and originally selected 250 candidate genes to test on **National Surgical Adjuvant Breast and Bowel Project (NSABP)** B-14 and B-20 trials. At the end they refined a 16+5 gene panel that could reliably predict recurrence (Mamounas *et al.*, 2010). The Mammaprint contains a 70-gene expression signature and the major trial in 2002 comprised women younger than 61 years, T1-T2, N0 disease (Van't Veer *et al.*, 2002; Van de Vijver *et al.*, 2002) The Prosigna from Nanostring contains a 50-gene expression signature plus five control genes and represents the old PAM50 assay. The PAM, or prediction analysis of microarrays, recapitulated the microarray classifier using RT-PCR-based PAM50 assay in comparison to standard clinical molecular markers. The major trial was in 2013 using a stage I–III cancer population and cleared by the Federal Drug Administration (FDA) in 2013. The PAM50 gene signature has been transferred to a novel and robust method for mRNA quantification (Van de Vijver *et al.*, 2002). This method works well in FFPE, does not rely on amplification of nucleic acids, and is intended for using kits in local labs with the proper instruments. The PAM50 expressions results are used to calculate a risk of recurrence score (ROR) and thus identify low-, intermediate-, and high-risk groups. The score is based on the intrinsic subtype and pathologic characteristics, with special weight given to a set of proliferation-associated genes. PAM50 correlates well with the Oncotype and the use of the four immunocytochemistry parameters (ER, PR, Her2, and ki67) (Dowsett *et al.*, 2013).

What we know for the diagnosis and prognosis of breast cancer, as described above, can be easily applied to other solid tumors and leukemias. Our knowledge and understanding of cancer is greater than a decade ago and we have a better perspective on the role of certain specific mutations in clue genes implicated in the disease. We also have many approved cancer drugs, some of which specifically target mutations in those genes. Oncologists are also getting better in the diagnosis as well as the initial treatment of the disease. Unfortunately, there are still many relevant challenges to face, such as drug resistance, tumor metastasis, the role of microenvironment, the immune response, the cancer-drug delivery system, and other areas that will be discussed in the next portion of this chapter and that need to be better understood by the new cadre of cancer researchers.

1.2.1. *Drug resistance*

Any cancer is a complex bio system that changes each time its cells divide. Thus the drugs that are used to kill cancers change not only the cancer cell but the whole microenvironment of the tumor. The selective pressure induced by the treatment activates protein pumps that work to get rid of the drug by changing the DNA repair of the cells or activating new signal transduction pathways. Ultimately the cancers create a resistance to the drugs that are supposed to kill them. The field of drug resistance is the new frontier in cancer research. Currently researchers are trying to understand at the molecular level what makes the cancer cell resistant to treatment that is intended to kill it. The identification of a signature that is unique to each tumor and or a subtype may help to better characterize the cancer cells that will be more susceptible to the killing effect of the therapy. Alternatively, identifying the cancer cell changes that occur with therapy through fluid biopsies using isolated cells from the blood or the bone marrow or the metastatic foci is a way to identify how the gene mutations are changing in the process of tumor progression. Comparing the cancer cells through their progression may help to determine how the cell evolves in its resistance to be killed. Intuitively the oncologists and experimental cancer researchers are aiming to overcome resistance either by increasing the doses or by using combinations of drugs that are targeted to different pathways. The main question is how to get the right combination as the number of genes that are mutated in cancer may be up to 300 but also can encompass the whole genome. The US National Cancer Institute (NCI) has been testing 5000 drug combinations against 60 cancer cell lines *in vitro*; the promising candidates are then screened for toxicity in mice. Yet even when this knowledge is finally obtained there will still be a lack of drugs that target tumor suppressor genes, and also areas of the genomic pathway, such as chromatin remodeling, splicing, and transcription process, which remain untargetable.

The task for the new cadre of cancer researchers is made even more challenging by the role of the microenvironment, which affects not only the cancer cells but also the normal cells. So how do the cancer cells escape the immune surveillance mechanisms? How do these events control cell dormancy and reactivation of cancer cells? These are significant biological questions that need to be answered before we dream of finding the right drug or combination of drugs to activate or abrogate these multiple processes that are all working simultaneously. Finding a drug to kill cells is probably not the kind of future that the new cancer researchers are focusing on, instead they must examine a more complex process of tissue remodeling that is taking place in cancer. This process must be borne in mind when cancer treatment is considered — not only in drug resistance. It could be that the drug resistance is only one of multiple targets that need to be considered. Areas like chromatin remodeling and manipulation of pluripotency are also aspects that must be also considered when studying drug resistance.

1.2.2. *Metastases*

Cancer patients rarely die due to their primary tumor, almost 90% of cancer deaths are a consequence of the metastatic process. It is true that we have a better understanding of both the mechanism of invasion and the metastatic process than a decade ago, but there is still a significant deficiency in our knowledge of the metastatic process, not least in our understanding of how the cell adapts to a new site or a distant tissue different from the primary tumor. Even more puzzling is how the circulating tumor cells are logged in the new site where some of them may remain dormant for years until they are reactivated. We need to understand how the initial spread of cancer cells takes place and how to control it by chemotherapy. Another intriguing question is what the switch that turns the cell metastatic is and how to recognize it in the primary tumor for prognosis of the cancer behavior or for determining the drug to be used for controlling this process? Even if we identify a drug that could be potentially beneficial in preclinical studies, it is difficult to detect this early event in cancer progression in a clinical trial. There is no doubt that new technologies will help researchers to have a better grasp of cell resistance and the metastatic process, but what technology is needed? Or is a change in the way that cancer researchers think on the gestalt of the biological process what is needed? Probably both processes go together and the innovative mind of the future cancer researcher must combine technological advances with an inner vision of creative thinking on how to approach cancer metastasis.

1.2.3. *Cancer metabolism*

The word "research" is a powerful one because it means to look or search again and again. In 1924, German biochemist Otto Warburg called the attention of the scientific community to his observation that cancer cells have a different metabolic pathway. Normal cells take glucose from the blood to generate heat and fulfill their tasks. Cancer cells do the same thing, but they burn much more glucose. Cells employ oxygen as oxidative phosphorylation and to burn glucose in their mitochondria using the tricarboxylic acid or TCA cycle or the Krebs cycle. The TCA cycle generates ATP, with carbon dioxide as a by-product, and when a cell becomes oxygen-depleted, or anaerobic, the cells are forced to switch to glycolysis. Compared with normal cellular metabolism, which can generate 36 units of ATP per molecule of glucose, glycolysis produces just two units, with lactate as a by-product. Warburg observed that the cancer cells use glycolysis even when they have enough oxygen. This aerobic glycolysis is now called the Warburg effect. Warburg proposed that this altered metabolism was the cause of cancer. The importance of glycolysis is that it provides energy for the formation of new proteins for receptors, new lipids for membranes, new nucleotides for DNA, and new fatty acids for cholesterol — all components required by the growing and dividing cells.

A further step in our knowledge was provided by the research work of Cantley (Rameh *et al.*, 1997) who discovered a molecular pathway known as PI3K, phosphoinositide 3 kinase, an enzyme that relays signals and activates other enzymes or genes. Cantley's group has shown that the PI3K signaling pathway is normally tightly controlled and in cancer that control is lost. The PI3K signaling pathway increases the activation of HIF-1a (hypoxia inducible factor 1 alpha) and produces more glucose transporters on the cell surface — both key changes that promote glycolysis in cancer cells. Because glycolysis happens so quickly it actually generates more ATP per second than the normal process does. Therefore, cancer cells using glycolysis also continue to metabolize glucose in their mitochondria, using the TCA cycle, to produce still more energy. That, in turn, generates free radicals — unstable molecules that can damage DNA, creating additional mutations that allow the activation of HIF-1a and produces more glucose transporters on the cell surface — both key changes that promote glycolysis in cancer cells (Klempner *et al.*, 2013). Glycolysis's waste product, lactic acid, might also make surrounding tissue more vulnerable to a tumor's infiltration. Because of this knowledge more than 100 clinical trials are based on the use of PI3K inhibitors.

What we know now is that cancer's metabolic reprogramming goes beyond glycolysis and the Warburg effect. Malignant cells also depend on glutamine, an amino acid that cells can metabolize to produce both energy and building blocks. Glutamine, unlike glucose, contains nitrogen, a key component of the nucleotides in DNA, RNA, and ribosomes, which make up much of the mass of a cell — and that all have to be duplicated when cells divide. It turns out that the oncogene Myc, essential for glycolysis, also directs cancer cells to scavenge glutamine out of the bloodstream. By inducing both glycolysis and glutamine metabolism, Myc uniquely supports the synthesis of the rapidly proliferating cells and tumors from early stage breast cancer patients, who had higher levels of glycine synthesis were more likely to die from the disease (see for more details, Corbet and Feron, 2015).

A mutated gene for a metabolic enzyme called IDH (isocitrate dehydrogenase), present in some leukemias, brain cancers, and gall bladder and gastrointestinal cancers, was shown to allow cells to synthesize an "oncometabolite" called 2-hydroxyglutarate (2HG), which modifies other enzymes, promotes the production of lipids and enhances tumor growth. IDH is mutated early in the development of those cancers, supporting the theory that cancer cells evolve toward altered metabolism, "selecting" mutated genes as they develop (Bogdanovic, 2015). The same reasoning is applied to pyruvate kinase (PK), which is considered a gatekeeper that helps determine whether a cell metabolizes glucose using the TCA cycle or diverts it to other pathways for anabolic growth. When cells become cancerous they use an isoform of PK, PKM2, in glycolysis instead, and Myc and HIF-1a, oncogenes that affect metabolism, may influence that switch. Because PKM2 is a common denominator in all cancers and clearly distinguishes them from many normal cells, it is a potential drug target (Wang *et al.*, 2015).

All these different discoveries on cancer cell metabolism are the main motif of the cancer process in terms of reprogramming behavior of these cells to grow and divide, if this cannot be accomplished the cancer cells enter in autophagy that is the consumption of the body's own tissue as a metabolic process. Autophagy is a new frontier for cancer researchers, controlling this process could create a new modality of cancer treatment (Dang, 2012; Vander Heiden, 2011).

It took almost 80 years to bring the revealing observations of Otto Warburg back to the research arena, and it seems that they will remain in the focus of scientific inquiry for quite a while. The new cadre of cancer researchers needs to consider the metabolomics of the cancer cells as a way to understand cancer's initiation and progression. Still more challenging is how other changes in the microbiome are affecting the systemic complexity of both normal and cancer cells, and also we need to think seriously on how the cancer cell may systemically affect the metabolic pathway of the whole organism.

1.2.4. *Immune surveillance and cancer*

Immunotherapy has been successful in cancer treatment when targeting B cells, which are white blood cells that become mutated in B-cell leukemia and lymphoma. During the past few years, promising results from early clinical trials have shown that therapies based on T-cell engineering can force some cancers into remission (Maude *et al.*, 2014). As these authors further elaborate the promise is that these immunotherapies are engineered for individual patients, using a person's own T cells to specifically target the cancer. The approaches are varied, but they all rely on the body's natural ability to distinguish its own substances from those of foreign intruders. When bacteria or viruses invade, some are swallowed by specialized antigen-presenting cells (APCs). These specialized cells are known as "dendritic cells" and were discovered by Ralph Steinman and described in the elegant paper on his identification of a novel cell type in peripheral lymphoid organs of mice and published in the *Journal of Experimental Medicine* in 1973. Steinman named these dendritic cells and their function is to chop up the bacterial or viral antigens and present them to T cells, white blood cells that orchestrate the immune response (Steinman and Cohn, 1973). The T cells then proliferate and launch a coordinated attack. The dendritic cells are not only responsible for activating T lymphocytes but also for orchestrating adaptive immune response toward microbes and tumor cells (Steinman and Bancherau, 2007). This means that the dendritic cells are the brain and executor of adaptive immunity. The intrinsic capacity of the dendritic cells to activate or silence the immune response according to the microenvironment or stimulus that they receive has been used for developing treatment strategies for prostatic, renal, and pancreatic cancer (Steinman and Banchereau, 2007).

As indicated by Cassiday (2014) new data are emerging from utilization of these basic concepts of immune response and the all gene therapy approaches. For example, gene therapy, the once-promising technique that uses engineered viruses and other methods to shuttle genes into human cells to fix DNA errors, faced a setback after an 18-year-old man died during a clinical trial in 1999. Later analysis showed that the virus carrying the DNA fixes had triggered a massive immune reaction that caused the man's organs to shut down. Gene therapies generally use viruses to deliver the gene of interest to cells. The viruses — retroviruses, lentiviruses and adeno-associated viruses, among others — are defanged by removing harmful genetic sequences then engineered to contain the therapeutic gene sequence. When the virus encounters human cells, it inserts its genetic material into the human genome. At present cancer researchers are aiming to develop vectors that would insert themselves into the genome more precisely, and thereby avoid disrupting important genes. In clinical trials for chronic lymphocytic leukemia (CLL) and acute lymphoblastic leukemia (ALL) there has been some success in isolating a patient's T cells and infecting them with a lentivirus that causes the cells to produce an antibody-like protein, or chimeric antigen receptor (CAR), on their surface, which binds to a protein on the surface of B cells. When the modified T cells were infused back into patients' bodies, they multiplied and attacked cancerous B cells. This development in immunotherapy is only the beginning of a more targeted approach in cancer treatment. The discovery of the CRISPR-cas 9 has opened even more interesting and innovative tools not only for immunotherapy but also for gene editing (see Chap. 8).

1.2.5. *Cell reprogramming*

In the human body, there are 60 trillion cells with around 200 different cell types. All of them originated from a unique cell, the fertilized ovocyte. This cell is a pluripotent cell that contains in its DNA the potential to develop an entire individual. It was Waddington (1957) who established the dynamic equilibrium between differentiated and undifferentiated cells, but it was Gurdon (1962) who demonstrated that an isolated nucleus of the cells of the intestine of a tadpole could be transplanted to an ovocyte that had previously had the nucleus eliminated and result in a complete tadpole. The creation of Dolly the sheep from the experiment of Wilmut, *et al.* (1997) was the final corroboration of the Gurdon discovery. Yamanaka's (2009) work introduced the concept that like human embryonic stem cells, or induced pluripotent stem (iPS) cells, could potentially be used as therapies, disease models, or in drug screening (Takahashi and Yamanaka, 2006). And iPS cells have clear advantages as the following: they can be made from adult cells, avoiding the contentious need for a human embryo, and they can be derived from people with diseases to create models or even therapies based on a person's genetic makeup (Itzhaki *et al.*, 2011; Zhao *et al.*, 2011). The originality of Yamanaka's works resided in the demonstration that four transcription factors Oct3/4; Sox 2, and Klf4 were enough to convert a mature fibroblast into a pluripotent cell (Takahashi and Yamanaka, 2006). From the start, biologists have tried to devise safer and more efficient recipes for making iPS cells than Yamanaka's method, which used a retrovirus to deliver a powerful shot of four genetic reprogramming factors into cells. Retroviruses integrate into a host cell's DNA and can therefore potentially disrupt gene expression and lead to cancer; and one of the reprogramming factors, Myc, is itself an oncogene that could cause cancer. The retroviral technique can transform about 0.01% of human skin stem cells into pluripotent cell lines; by comparison, adenoviruses, which do not integrate into the genome, transform just 0.0001–0.0018% of cells (González *et al.*, 2011], and delivering the reprogramming factors directly into a human cell can transform 0.001% of cells. Therefore, it is expected that more work in this area will bring this basic knowledge of developmental biology to the treatment of cancer patients by reprogramming the cancer cells to a normal differentiated phenotype.

There are significant questions to ponder, such as the use of CRISPR-Cas 9 for editing the genome (see Chap. 8) but this is definitely one of the topics that the new cadre of cancer researchers will need to face. For example, CRlsPR-Cas9 has been used for understanding the role of loops of DNA called "insulated neighborhoods" that can protect small groups of genes from silencing or activation. These loops of DNA are helping to maintain the DNA packed into the nucleus with a great precision for controlling gene expression. It has been postulated (Hnisz *et al.*, 2016) that releasing DNA loops allows physical contact between proto-oncogenes and nearby enhancers. CRIsPR-Cas9-mediated deletions of insulated neighborhood boundaries can activate proto-oncogenes. Although this mechanism needs to be further studied, it is pointing to the importance of the 3D architecture of the genome not only in normal but also in cancer cells.

Reprogramming the cancer cells also requires knowledge of as to how cell reprograming is controlled by a complex network of proteins, chemical modifications, and RNA that together organize genomic DNA into chromatin. In a recent publication in *Science*, it is provided a model of "algorithmic" chromatin biology, or unifying epigenetic regulation (Bintu *et al.*, 2016). In this publication, the authors proposed the 3-state model of gene regulation by repressive chromatin regulators, in which cells stochastically transition between active, reversibly silent, and irreversibly silent states. Basically they have studied individual cells during chromatin regulators recruitment, and reactivation events upon release of the chromatin regulators, by means of time-lapse microscopy. Although this publication does not solve all the intricacy of chromatin remodeling, it provides an important initial step that future cancer researchers can develop.

1.2.6. *Drug delivery*

Chemotherapy drugs kill rapidly dividing cells to prevent the rampant cell growth that results in tumors. But these drugs reach cancer cells through the same network of blood vessels that supplies the whole body, According to Wright (2014), building protective coats around toxic molecules could address one of cancer treatment's biggest remaining challenges: how to spare healthy cells when attacking cancerous ones. Therefore, how to shuttle the drug to the target cells is one of the areas that offers significant promises as well as challenges in cancer research. The drug delivery is a multi-stage journey that starts when the active agent needs to enter the body, travel through the bloodstream to arrive at the tumor site, penetrate the tumor mass and then gain entry to the cells. Developments in cancer therapy have focused on drugs that target cancer-specific biological pathways. One of the simplest ways to do this is to arm cancer-seeking proteins with a cell-killing drug. In the late 1990s, drug companies developed anti-bodies that bind to the surface of certain types of cancer cells. Fusing these proteins to drugs with a stable chemical linker yields a potent combination: the antibodies encourage uptake of the drug into cancer cells and the linker keeps the drug from working until it gets inside. Only two antibody–drug conjugate (ADC) therapies are on the market. The first, called "bren-tuximab vedotin" (Seattle Genetics in Washington), was approved by the US Food and Drug Administration (FDA) in 2011 to treat some types of lymphoma that had not responded to previous treatment. The second, "trastuzumab emtansine" (marketed by Genentech of San Francisco, California) was approved in 2013 as a treatment for late-stage breast cancer after treatment with conventional chemotherapeutics. Trastuzumab emtansine fuses emtansine, a toxic chemotherapeutic, to antibodies that bind to a protein receptor called HER2, which is overproduced by about 20% of breast cancers. In a phase III trial that finished in 2012, the nearly 500 women who took the drug lived about 5 months longer and had fewer side effects than did those on the standard treatment. Several similar drugs are now in advanced clinical trials.

Besides antibodies, another approach is to add a coat around chemotherapy drugs. The resulting nanoparticles, which range from about 20–100 nanometers in diameter, are too large to escape most blood vessels. But they do find their way out of the leakier ones hastily built by a rapidly growing tumor. As a result, they are thought to accumulate preferentially at the tumor site. Nanoparticles can carry a stronger payload than can antibodies, encasing thousands of drugs in a single molecule. The problem is the secondary effects of accumulation in the liver and the spleen, where nanoparticles provide no therapeutic benefit. To minimize unwanted effects the particles are coated with a layer of polyethylene glycol; this mimics water and effectively hides the drug from the liver cells that detect and engulf intruders. The first nanoparticles to be developed for drug delivery coated the active agent with lipids. The first drug of this type was Doxil, which carries doxorubicin and is used to treat Kaposi's sarcoma and other solid tumors, including breast and ovarian cancers. According to the US National Cancer Institute, six such nanoparticles are currently approved for use on the market worldwide. So far, they seem to improve safety but not efficacy. Because the cost is 10 times more than conventional treatment, some are questioning whether nanotechnology is worth the high price tag that accompanies its production and that is yet another question that the new cadre of cancer research need to face and solve.

Despite these setbacks, there are new ways to look at the problem and using nanoparticles to deliver small pieces of RNA to cancer cells using basically RNA interference is promising. The advantage of this approach is that the use of nanoparticles to deliver RNA has the potential to reach multiple genes, and thus pathways, in one hit. For example, the particle, called CALAA-01, targets the gene RRM2, which is involved in cell division and uses molecules that bind to the transferrin receptor, which is highly expressed on cancer cells, to gain access to the cell interior of melanoma cells. Probably the use of Interference RNA will be overshadowed by the use of CRISPR-cas9 technology (See Chap. 8).

According to Verma *et al.* (2012), another effort to deliver drugs directly to the tumor cells involves building magnetic particles that will be able to be target the cancer. These magnetic particles are similar to the contrast agent used in magnetic resonance imaging (MRI), so when a patient is in the MRI scanner the strong magnetic field on top of the cancer mass can be used to guide drugs to the correct site. This method has been tested in pigs but not yet used in humans. However it is innovative enough to offer promise in the drug delivery system.

1.3. The Challenge of Individual Versus Common Good in Personalized Medicine

A critical area of current debate is the use of personalized medicine in a social environment, which involves a collision between individual rights and the common good. Cressey reported that in Great Britain there are concerns that patients in the National Health Service were not being given the most effective treatments created the National Institute for Health and Clinical Excellence (NICE). The purpose is to make sure that the National Health Service spends its budget using a transparent decision-making process that is based on the best evidence available. However even if one treatment is more effective clinically than another, NICE committee members can decide that the gain in health that it bestows is not worth the additional cost. The influence of this idea is reaching other countries including Azerbaijan and Brazil, which have adapted the institute's guidance. The United States has no such limit on health-care spending. The nation's Food and Drug Administration approves a treatment for use if it is deemed to be both safe and effective for the licensed condition — cost is not a consideration. "The American health-care system rations on the basis of whether people have private insurance or not" (Cressey, 2009). On the basis of these facts, many critical questions emerge, such as: Who will decide the fate of the individual cancer patient? How will the concept of personalized medicine be applied and who will decide the payment of the procedure? Are we expecting that all individuals will have equal insurance that will cover all the possible variations in treatment for each individual cancer? These are real problems that the new cadre of cancer researchers will need to face at the end. In the next few sections I would like to dissect part of these problems, but they are not the only factors for consideration.

1.3.1. *Genetic testing*

Genetic testing is one part of oncology practice in which researchers will continue to have an input into its development. According to Vogelstein *et al.* (2013), the analysis of tumor genomes has revealed some 140 genes whose mutations contribute to cancer. The problem is that although a significant number of cancer patients are getting their DNA sequenced, the use of these data in a personalized medicine approach lags behind. There are fewer than 60 genetic variants that are deemed worthy for use in clinical care and the most frequently used are the variants of BCRA and EGFR genes. Mutations in the BRCA genes have about an 80% chance of developing breast cancer, leading some who carry this mutation to opt for preventive mastectomies. Mutations in the EGFR gene can indicate whether lung cancer will respond to expensive drugs with fewer side effects than standard chemotherapy (Vogelstein *et al.*, 2013). According to some reports (Lynch *et al.*, 2013), 5 years after EGFR tests were commercialized, only around 6% of appropriate US patients were being genotyped, partly because their physicians were unaware of the tests. It has been estimated that the clinical-sequencing market is more than US$2 billion; therefore for these investments to pay off, there needs to be an efficient way to evaluate whether genetic information leads to better health care. According to Lynch *et al.*, where the predictions are in target and the knowledge is there, what is really missing is the evidence that each treatment produces the desired outcome, meaning we must have better information on the clinical response among the mutation, the treatment, and the response of the patient to the putative treatment.

There are promising stories wherein the DNA typing of tumors suggests clear approaches to therapy, with improved results for patients. The basic principle is that if the cancer researchers and clinicians find the key genetic mutations that drive a particular cancer's growth, they will be able to target the tumor more selectively and with fewer toxic side effects. However we do not know enough about which genetic mutations drive a given cancer, let alone how to interrupt the aberrant cellular pathways that result in cancer progression (Gravitz, 2014). The effectiveness of imatinib against chronic myelogenous leukemia (CML) offers hope for the future, but this success has not been easy to duplicate. The explanation for not finding the right drug is that every tumor has a unique set of genetic mutations, which makes it difficult to match a patient with the appropriate therapy. This reality of cancer biology has an important proposition, namely to classify tumors by their mutations and expression profiles as opposed to the way they look under the microscope. There are some studies that are already using this approach (Gravitz, 2014). For example, mutations in the gene BRAF can cause the protein it encodes to become oncogenic and a drug called vemurafenib, which is effective against melanomas that contain a mutation in the BRAF protein, has been used in patients with other types of cancer who test positive for the same mutation. Although the equation seems to be obvious, there are realities of cancer biology that are of concern because treating cancer by addressing a mutation in a cell that is continuously mutating could be only part of the problem. Another issue with the targeted approach is that therapies are typically tested only on patients with advanced cancers, which are much harder to treat than those in their early stages. But perhaps the biggest obstacle to targeting the products of mutated genes is that so many of the causative mutations result not in something's presence but in its absence, such as tumor suppressor genes. The Gravitz's article speculates "that precision medicine also depends on providing the drugs at the right time and this requires to know not only which mutations got a tumor started, but also how the tumor is likely to change". Technology can help provide a more dynamic way to understand how these changes are taking place by using liquid biopsies of cancer circulating cells in the blood. Using liquid biopsies cancers can be genetically characterized through the utilization of genome-sequencing techniques by analyzing DNA taken from a blood sample. Liquid biopsies can study the DNA of circulating tumor cells; or the whole tumor cells in the bloodstream and also by capturing small vesicles called "exosomes" that are ejected by tumor cells. The advantages of the liquid biopsies are several, among them that they are easy to obtain, allow a fast monitoring of therapy response and in some cases may predict tumor response and diagnosis of cancer before they are clinically present. The possibility to perform DNA analysis in the circulating cells or in the free DNA could allow the identification of cancer genes in the body. The best utilization of DNA sequencing and digital polymerase chain reaction (dPCR) allows researchers to detect or quantify specific stretches of tumor DNA even at very low concentration. Small DNA fragments floating in the bloodstream are also derived from normal cells; however, they are around 100–200 base pairs long, and are still stacked to histones and because this packaging of DNA by histone is tissue specific, it could help to determine normal from cancer-originated DNA. The identification of circulating DNA in cancer patients can also provide a clue on recurrence of a disease, presumably predicting who will be free of disease. A new finding is that platelets in the blood may contain vesicles loaded with tumor cells RNA providing a new way to look at the liquid biopsies. Although liquid biopsies are extremely promising, more testing and validation needs to be done by the new cadre of cancer researchers before it is finally implemented in the clinical armamentarium.

1.3.2. *The reality of genetic testing*

Targeting a personalized therapy is a problem that cancer researchers are facing, and it is not an easy one because it requires the existence of a global system of clinical data that can be referenced by the patient record and compared with the response. According to Ginsburg (2014), what is needed is a health-care system with comprehensive electronic medical records and accessible stores of patient data. There are isolated examples aiming to solve this problem but not a global system. The US National Human Genome Research Institute is exploring the clinical implementation of genomic information through the Electronic Medical Records and Genomics (eMERGE) network, a consortium of nine institutions that use commercial and homegrown electronic medical records to capture genomic information, store it securely and incorporate it into computer algorithms to guide clinicians. The idea behind this project is that it will allow clinicians to study the outcome of treatments guided by genetic information compared with usual care. The basic problem, according to Ginsburg, is "that most of the existing patient databases for clinical care are often inadequate for evaluating health interventions: data are frequently missing or incorrect, and it can be hard to link data about, e.g., demographics, appointments, procedures, medications, and vital signs". Quackenbush's (2014) article in *Nature* presents a lucid analysis of the patient data collection necessary to deliver actionable point-of-care information to doctors treating patients. He cites the "challenge to develop analytical methods that are effective for huge amounts of genomic data. In particular, better methods are needed to 'normalize' the data generated by different technologies or at different sites, so that results can be compared across studies". Quackenbush insisted that "the greatest barrier to the use of "big data" in biomedical research is not one of methodology. It is, rather, the lack of uniform, anonymized clinical data about the patients whose samples are being analyzed. Unfortunately, nearly every published study lacks the clinical data to address fundamental research questions fully or to allow the findings of one study to be validated in others". Quackenbush has three main suggestions: (1) develop more flexible patient-consent procedures; (2) develop hospital and laboratory computational-infrastructure and data-security protocols to improve the sharing, access, and fair use of clinical data; and (3) change the culture of data sharing. In the last point, he emphasized that the sharing of clinical data in most publications dealing with genomic data is frequently limited to a bare minimum. Quackenbush stated "that the absence of such key information again makes it difficult to reproduce the results of an analysis or to validate other published data sets. Big data has tremendous potential to provide fresh insights into diseases such as cancer. But that potential will be realized only by tackling how best to share the clinical information necessary to interpret it. And developing a more complete understanding is essential if we are ultimately to create the knowledge base necessary to provide clear, concise, reliable, and actionable information to doctors and their patients".

Present and future cancer researchers must fix these problems, and that will require policy makers and administrators in the health-care system to review the existing way in which data are collected and stored. There is some progress at the societal level, recently a US Supreme Court ruling invalidated the patenting of genes, and with it certain companies like Myriad's exclusive rights on two genes associated with breast- and ovarian-cancer risk. However the firm still has a private trove of data from 1.3 million genetic tests, giving them an advantage in interpreting test results on these genes. New players with a different view but a common goal are now emerging to make their individual information part of a central database that counteracts the monopolies held by companies, says Hayden (2014). This data sharing has also been facilitated by reducing the cost of genomic testing, and a fair share of the credit for this success goes to a grant scheme run by the US National Human Genome Research Institute (NHGRI). Officially called the Advanced Sequencing Technology awards, it is known more widely as the $1,000 and $100,000 genome programs. Started in 2004, the scheme has awarded grants to several groups of academic and industrial scientists, including some at every major sequencing company and the fruit of this cooperative effort is nearly ready (Hayden, 2014).

1.4. Training the New Cadre of Cancer Researchers

Right now we need to face the all-important question of how the new cadre of cancer researchers will be trained so they can make the right questions that will help them to solve the problem of cancer. What kind of training does the new generation of scientists require to make this difference? In a time in which big science is on the spot and multiple collaborations between different investigators, institutions, and countries is the norm, how will the individual mind find a way to make a difference? It is my strong belief that only a few will have the natural acumen to see the whole picture and solve the problem or problems, although perhaps those will be the ones who establish the teams and groups that eventually help to solve the problem of cancer. Without a doubt, somebody will need to be the conductor in this complex orchestra, and there are several criteria that will be critical in the formation of the new cadre of cancer researchers.

1.4.1 *Sense of history*

The accumulation of our scientific knowledge probably began in early 16 century by using the same system that the lawyers have been using that was the accumulation of facts. Lawyers who accumulated these facts had to establish agreed-upon observations as evidence that could not shift later. Modern science is the result of the accumulation of these facts and the cancer research endeavor is still accumulating knowledge that can be quantitated and reproduced in order to find common paths for treatment and prevention. As cancer researchers, we must be aware that our contributions are part of a continuous chain of advances in the historical framework of the time. For example, in 1592 Francis Bacon established that the correct way to experiment is to change one variable at a time. Cancer research literature demonstrates and repeats experiments to show reproducibility in the outcome of our approach that was established by Robert Boyle in the early seventeen century. The same reasoning applies to our modern way to evaluate our research by submitting to referee-reviewed journals that add a layer of confirmation and validation over shared knowledge. The concept of blinded, randomized design did not emerge until late 19th century and the double-blind experiment is a recent development of the 1950s. The meta-analysis, a second level of analysis of all previous analyses in a given field, was created in 1974. These examples are an indication of the rapid pace at which scientific knowledge is acquired. The new cadre of cancer researchers will be in the vortex of this process. The number of journals and publications is increasing every year and to keep updated in the field, even in the narrowest specialties of cancer research such as kinase inhibitors, is a major daily task. How to parse the literature and sort those data that are validated from those that are only initial ideas? How to distinguish negative data from those that are published in the less illustrious journals? Or even how to certain that those data published in prestigious journals will not be recalled because the data are not as original as previously represented. The most difficult question is whether the peer system of journal publication will be maintained along with the pressure to make everything transparent and easily accessible by everyone? Will the new cadre of cancer researchers face a system in which the data will be available to them even in the early phases of the research process? I cannot answer all these questions but they are things to come. This could be a science fiction as depicted in Dave Eggers's book, *The Circle*, but what is described in that book as a fictional story is only a mirror image of what is already going on in our social media. This growing need of our millennial generation to know and be known has created a system in which contemplation and thinking as an individual have been transformed into difficult tasks. However, I believe in the resourcefulness of our species; we will find new ways of creative thinking.

Another area of concern is how will the new cadre of cancer researchers keep abreast of technological advances? Where will the funds for new equipment come from? Will the RO1 system of grants — which is the last stand for independent research — be maintained? And how will scientific research institutions be able to provide the resources needed to keep pace with the demands of technological advances?

In the middle of this whirlpool of technological advances and information will be the individual researcher who needs to find a way through this historical process of changes and advances that we have never before seen. Keeping a perspective in history of our humanness will be an important quality of the members of this new cadre of cancer researchers.

1.4.2. *Seeing the problem in nonconventional ways*

The ability to examine the biological problem from new angles and in nonconventional ways is another important quality that the new cadre of researchers will need to develop and preserve, although this is also a responsibility of the present generation of cancer researchers who must listen to and foster new approaches if they expect changes in their lifetime.

Fostering independence requires good mentors and an academic environment that nurtures this process. A practical problem is the funding of the research endeavor that is flexible enough to facilitate this process.

Although many granting agencies are fostering these kinds of innovative ideas, the present peer-review system is not as open minded as we would expect it to be and good potential ideas are not funded indicating that our academic environment is not as open as it needs to be for the new cadre of cancer researchers to succeed. Therefore, it is the responsibility of our present generation of cancer researchers to create an adequate academic environment that fosters flexible projects and stimulates as well as trains the new cadre by offering the ability to manage more than one research project at the same time. No cancer researcher of our generation knows what is coming or what potential paths will emerge. Therefore, it is the responsibility of the present generation of cancer researchers to make the focus of the new generation's research more open and flexible.

1.4.3. *Thinking and performing cancer research endeavors in a societal framework*

There is no doubt that those members of the new cadre of cancer researchers who are able to combine all the new advances of the genomics, proteomics, epigenomics, and metabolomics with mathematical modeling and biological processes will have the best chances to be the driving forces in cancer research. But in addition to these qualifications the new cadre of cancer researchers will need to have a good understanding of the minority and disparities issues, not only in the research endeavor but also in the application of this knowledge to the patient — meaning the social component of their work. Issues like disabilities, genre, and economic disparity must be somehow engraved in the social consciousness of the cancer researcher. For example, people with disabilities have a consistently higher unemployment rate than the general population for obvious reasons (Gewin, 2011). If the cancer researcher is one of those people with a disability, their challenge to go ahead is not an easy one. Therefore, the granting agencies need to consider special funding and supplements for them.

The new cadre of cancer researchers needs to learn how to deal not only with hard core research and the translation to the clinical field but they also need to have a perception of the ongoing trends in our society, such as diet issues, exercise, and cultural differences. The more researchers learn how genes, proteins, and metabolic pathways interact with diet, exercise, and cultural or ethnic differences, the more they will be able to use an interdisciplinary approach in cancer research. Therefore, the more researchers understand epidemiological trends and also employ intervention studies — which are more like the clinical trials used by drug and medical device makers — the more translational and rooted in the reality of the human condition the cancer research will be (Laursen, 2010).

This translational view of cancer research applies to the ability of the new cadre of cancer researchers to communicate ideas. In a visual society like we live in, ideas are transmitted through images and movies making the point more efficiently than any other media. Although there are agencies like the Science and Entertainment Exchange of the US National Academy of Sciences — which links scientists and engineers with movie and television-show makers to provide, "the credibility and verisimilitude on which quality entertainment depends" (Sarewitz, 2010; Savage, 2014; Serageldin, 2011) and although this has been helping to improve the perception of science in our society, still we are far away from perfection. Therefore, it is our responsibility as scientists to ensure that the message of science, mainly in cancer, is properly transmitted.

References

Adams, J. The fourth age of research. *Nature,* **497**: 557–560, 2011.

Balkwill, F. and Mantovani, A. Inflammation and cancer: Back to Virchow. *Lancet,* **357**: 539–545, 2001.

Bintu, L., Yong, J., Antebi, Y.E., McCue, K., Kazuki, Y., Uno, N., Oshimura, M., and Elowitz, M.B. Dynamics of epigenetic regulation at the single-cell level. *Science,* **351**: 720–724, 2016.

Bloom, H.J.G. and Richardson, W.W. Histologic grading and prognosis in breast cancer: A study of 1409 cases of which 359 have been followed for 15 years. *Br. J. Cancer,* **11**: 359–377, 1957.

Bloom, H.J.G., Richardson, W.W., and Field, J.R. Host resistance and survival in carcinoma of breast: Study of 104 cases of medullary carcinoma in a series of 1411 cases of breast cancer followed for 20 years. *Br. Med. J.* **3**: 181–188, 1970.

Bogdanovic, E. IDH1, lipid metabolism and cancer: Shedding new light on old ideas. *Biochim. Biophys. Acta.* **1850**: 1781–1785, 2015.

Bradshaw, R.A. Rita Levi-Montalcini (1909–2012). *Nature,* **493**: 221, 2013.

Caskey, T.C. Marshall Nirenberg (1927–2010). *Nature,* **464**: 44, 2011.

Cassiday, L. Gene-therapy reboot. *Nature,* **509**: 651–652, 2014.

Corbet, C. and Feron, O. Metabolic and mind shifts: From glucose to glutamine and acetate addictions in cancer. *Curr. Opin. Clin. Nutr. Metab. Care,* **18**(4): 346–353, 2015.

Cressey, D. Health economics: Life in the balance. *Nature,* **461**: 336–339, 2009.

Dang, C.V. Cancer cell metabolism: There is no ROS for the weary. *Cancer Discov.* **2**: 304–307, 2012.

Dowsett, M., Sestak, I., Lopez-Knowles, E., Sidhu, K., Dunbier, A.K., Cowens, J.W., Ferree, S., Storhoff, J., Schaper, C., and Cuzick, J. Comparison of PAM50 risk of recurrence score with oncotype DX and IHC4 for predicting risk of distant recurrence after endocrine therapy. *J. Clin. Oncol.* **31**: 2783–2790, 2013.

Gewin, V. The fight for access. *Nature,* **469**: 257, 2011.

Ginsburg, G. Gather and use of genetic data in health care. *Nature,* **508**: 451–453, 2014.

González, F., Boué, S., and Izpisúa Belmonte, J.C. Methods for making induced pluripotent stem cells: reprogramming à la carte. *Nat. Rev. Genet.* **12**: 231–242.

Gravitz, L. This time it's personal. *Nature,* **509**: S52–S54, 2014.

Gurdon, J.B. The developmental capacity of nuclei taken from intestinal epithelium cell of feeding tadpoles. *J. Embryol. Exp. Morphol.* **10**: 622–649, 1962.

Hanahan, D. and Weinberg, R.A. Hallmarks of cancer, the next generation. *Science,* **144**: 646–654, 2011.

Harris, H. Concerning the origin of malignant tumors by Theodor Boveri. Translated and annotated by Henry Harris. *J. Cell Sci.* **121**(Suppl 1): 3, 2008.

Hayden, E.C. Cancer gene data sharing boosted. *Nature,* **510**: 198, 2014.

Hnisz, D., Weintraub, A.S., Day, D.S., Valton, A.L., Bak, R.O., Li, C.H., Goldmann, J., Lajoie, B.R., Fan, Z.P., Sigova, A.A., Reddy, J., Borges-Rivera, D., Lee, T.I., Jaenisch, R., Porteus, M.H., Dekker, J., and Young, R.A. Activation of proto-oncogenes by disruption of chromosome neighborhoods. *Science.* **351**: 1454–1458, 2016.

Itzhaki, I., Maizels, L., Huber, I., Zwi-Dantsis, L., Caspi, O., Winterstern, A., Feldman, O., Gepstein, A., Arbel, G., Hammerman, H., Boulos, M., and Gepstein, L. Modelling the long QT syndrome with induced pluripotent stem cells. *Nature,* **471**: 225–229, 2011.

Klempner, S.J., Myers, A.P., and Cantley, L.C. What a tangled web we weave: Emerging resistance mechanisms to inhibition of the phosphoinositide 3-kinase pathway. *Cancer Discov.* **3**: 1345–1354, 2013.

Laursen, L. Big science at the table. *Nature,* **468**: 52–54, 2010.

Lynch, J.A., Khoury, M.J., Borzecki, A., Cromwell, J., Hayman, L.L., Ponte, P.R., Miller, G.A., and Lathan, C.S. Utilization of epidermal growth factor receptor

(EGFR) testing in the United States: A case study of T3 translational research. *Genet. Med.* **15**: 630–638, 2013.

Mamounas, E.P., Tang, G., Fisher, B., Paik, S., Shak, S., Costantino, J.P., Watson, D., Geyer, C.E. Jr., Wickerham, D.L., and Wolmark, N. Association between the 21-gene recurrence score assay and risk of locoregional recurrence in node-negative, estrogen receptor-positive breast cancer: Results from NSABP B-14 and NSABP B-20. *J. Clin. Oncol.* 281677–281683, 2010.

Maude, S.L., Frey, N., Shaw, P.A., Aplenc, R., Barrett, D.M., Bunin, N.J., Chew, A., Gonzalez, V.E., Zheng, Z., Lacey, S.F., Mahnke, Y.D., Melenhorst, J.J., Rheingold, S.R., Shen, A., Teachey, D.T., Levine, B.L., June, C.H., Porter, D.L., and Grupp, S.A. Chimeric antigen receptor T cells for sustained remissions in leukemia. *N. Engl. J. Med.* **371**: 1507–1517, 2014.

Mukherjee, S. *The Emperor of All Maladies: A Biography of Cancer*, Scriebner: New York, 2010.

Pederson, T. and Paul C. Zamecnik (1912–2009): Trailblazer in the study of protein synthesis. *Nature*, **462**: 413, 2009.

Quackenbush, J. Learning to share. *Nature*, **509**: 568, 2014.

Rameh, L.E., Tolias, K.F., Duckworth, B.C., Cantley, L.C. A new pathway for synthesis of phosphatidylinositol-4, 5-bisphosphate. *Nature*, **390**: 192–196, 1997.

Russo, J. *The apprentice of science: A handbook for the budding biomedical researchers*. London: World Scientific Publishing Co, 2010.

Russo, J., Frederick, J., Ownby, H.E., Fine, G., Husain, M., Krickstein, H.I., Robbins, T.O., and Rosenberg, B.F. Predictors of recurrence and survival of breast cancer patients. *Am. J. Clin. Pathol.* **88**: 132–138, 1987.

Sarewitz, D. Entertaining science. *Nature*, **466**: 27–28, 2010.

Savage, N. Big data versus the big. *Nature*, **509**: S67, 2014.

Sharma, P., Allison, J.P. Lloyd J. Old (1933–2011). *Science*, **335**: 49, 2012.

Serageldin, I. The values of science. *Science*, **332**: 1127, 2011.

Sørlie, T., Perou, C.M., Fan, C., Geisler, S., Aas, T., Nobel, A., Anker, G., Akslen, L.A., Botstein, D., Børresen-Dale, A.L., and Lønning, P.E. Gene expression profiles do not consistently predict the clinical treatment response in locally advanced breast cancer. *Mol. Cancer Ther.* **5**: 2914–2918, 2006.

Steinman, R.M. Linking innate to adaptive immunity through dendritic cells. Novartis Found. *Symposium*, **279**: 101–109, 2006.

Steinman, R.M. and Bancherau, J. Taking dendritic cells into medicine. *Nature*, **440**: 419–426, 2007.

Steinman, R.M. and Cohn, Z.A. Identification of a novel cell type in peripheral lymphoid organs of mice, I, Morphology, quantitation, tissue distribution. *J. Exp. Med.* **137**: 1142–1162, 1973.

Takahashi, K. and Yamanaka, S. Induction of pluripotent stem cells from mouse embryonic and adult fibroblasts cultures by defined factors. *Cell,* **126**: 663–676, 2006.

Titford, M. Rudolph Virchow: Cellular pathologist. *Lab Med.* **41**: 311–312, 2010.

Vander Heiden M.G. Targeting cancer metabolism: A therapeutic window opens. *Nat. Rev. Drug Discov.* **21**: 6322–6327, 2011.

Van't Veer, L.J., Dai, H., Marton, M.J., Witteveen, A.T., Schreiber, G.J., Kerkhoven, R.M., Roberts, C., Linsley, P.S., Bernards, R., and Friend, S. H. Gene expression profiling predicts clinical outcome of breast cancer. *Nature,* **415**: 530–536, 2002.

van de Vijver, M.J., He, Y.D., Van't Veer, L.J., Dai, H., Hart, A.A., Voskuil, D.W., Schreiber, G.J., Peterse, J.L., Roberts, C., Marton, M.J., Parrish, M., Atsma, D., Witteveen, A., Glas, A., Delahaye, L., van der Velde, T., Bartelink, H., Rodenhuis, S., Rutgers, E.T., Friend, S.H., and Bernards, R. A gene-expression signature as a predictor of survival in breast cancer. *N. Engl. J. Med.* **347**: 1999–2009, 2002.

Venitt, S., Phillips, D.H. and Philip D. Lawley (1927–2011): Chemist who discovered that cancer is caused by damage to DNA. *Nature,* **482**: 36, 2012.

Verma, S., Miles, D., Gianni, L., Krop, I.E., Welslau, M., Baselga, J., Pegram, M., Oh, D.Y., Diéras, V., Guardino, E., Fang, L., Lu, M.W., Olsen, S., and Blackwell, K. EMILIA study group trastuzumab emtansine for HER2-positive advanced breast cancer. *N. Engl. J. Med.* **367**: 1783–1791, 2012.

Vogelstein, B., Papadopoulos, N., Velculescu, V.E., Zhou, S., Diaz, L.A. Jr., and Kinzler, K.W. Cancer genome landscapes. *Science,* **339**: 1546–1558, 2013.

Waddington, C.H. *The Strategy of the Genes: A Discussion of Some Aspects of Theoretical Biology.* London: Allen & Unwin, 1957.

Wang, Y., Zhang, X., Zhang, Y., Zhu, Y., Yuan, C., Qi, B., Zhang, W., Wang, D., Ding, X., Wu, H., and Cheng, J. Overexpression of pyruvate kinase M2 associates with aggressive clinicopathological features and unfavorable prognosis in oral squamous cell carcinoma. *Cancer Biol. Ther.* **13**: 1–7, 2015.

Wilmut, I., Schnieke, A.E., McWhir, J., Kind, A.J., and Campbell, K.H.S. Viable offspring derived from fetal and adult mammalian cells. *Nature,* **385**: 810–813, 1997.

Wright, J. Deliver on a promise. *Nature,* **509**: S59, 2014.

Yamanaka, S. Ekiden to iPS cells. *Nat. Med.* **15**: 1145–1148, 2009.

Zhao, T., Zhang, Z.N., Rong, Z., and Xu, Y. Immunogenicity of induced pluripotent stem cells. *Nature,* **474**: 212–215, 2011.

Further Reading

Baltimore, D. Retrospective Renato Dulbecco (1914–2012). *Science*, **335**: 1587, 2012.

Bernstein, R. Cancer killers. *Nature*, **508**: 139–140, 2014.

Tang, G., Shak, S., Paik, S., Anderson, S.J., Costantino, J.P., Geyer, C.E. Jr., Mamounas, E.P., Wickerham, D.L., and Wolmark, N. Comparison of the prognostic and predictive utilities of the 21-gene Recurrence Score assay and Adjuvant! for women with node-negative, ER-positive breast cancer: results from NSABP B-14 and NSABP B-20. *Breast Cancer Res. Treat.* **127**: 133–142, 2011.

Wright, J.R. and Albert, C. Broders' paradigm shifts involving the prognostication and definition of cancer. *Arch. Pathol. Lab. Med.* **136**: 1437–1446, 2012.

Chapter 2

Trends in Scientific Discovery

2.1. Introduction

Although cancer has been with us since the first humans walked the Earth, it is actually not one disease but many, depending on the cell of origin and its genotype. Many of the genes involved in cancer are important in cell communication, growth signaling, apoptosis, and DNA repair, and when these functions are disrupted in a cell, cancer can result. Basically cancer is a consequence of mutations in the genes that control the basic process of life, such as oncogenes and tumor-suppressor genes. Mutations in these genes are induced by chemicals, radiation, or viruses that may up-regulate or down-regulate a gene. Expanding knowledge of the human genome is helping our understanding of what goes wrong with genes in cancer tissue. Therefore it seems that the paths are now open for the new cadre of cancer researchers to not only better understand what cancer is all about, but also to treat and prevent the disease, in all its forms. However, the knowledge in front of us is a complex network of thousands of interactions between molecules, and a true understanding of cells and organisms will require contributions from mathematicians, computer scientists, engineers, physicists, and chemists to complement the efforts of biologists. Furthermore, the first issue that cancer researchers must face is selecting which problem could be most effectively investigated with the tools that he or she has available. The short-term goal is to generate creative knowledge that makes the field of cancer move forward, and long term we are basically talking about a real-breakthrough. For this to happen researchers will need to frame the right question, then identify which scientists have the know-how and the necessary supporting network of resources, including researchers and institutions that break barriers among individuals.

Scientific excellence is probably the first requirement, but it must be matched with diversity in disciplines, and good communication among these varied individuals will be a must. In the next sections of this chapter, I discuss the major fields of research that the new cadre of cancer researchers must explore.

2.2. Controlling Chromatin Remodeling

Chromosomes are DNA-containing structures coiled in the nucleus of eukaryotes. Two meters of DNA are packed in spheres not more than 10 μm across; these spheres are cell specific and deeply related to gene transcription. We know that DNA double helices coil around proteins called histones, forming chromatin strands that are, in turn, bundled into chromosomes. How these strains of DNA interact with each other is still in the process of being elucidated. There is a long-range looping interaction that allows gene sequences to physically contact regulatory elements that are farther away (Carter *et al.*, 2002). In many cancers, this interaction between different parts of the chromosomes explains chromosome translocations because these events occur more often between genes that physically come together during transcription. Chromatin structure has become the focus of the new cadre of cancer researchers because of the discovery of noncoding areas of DNA that play a more important role in gene transcription that previously imagined (Santucci-Pereira *et al.*, 2014; Barton *et al.*, 2014).

It is now recognized that DNA sequences and histones are epigenetically modified, and so the key to turning on and off gene transcription and a prime new research area is in understanding the ways that chromatin folds, moves, and communicates. Thus new concepts such as chromatin network, chromosome interactome, and spatial epigenetics are now the focus of cancer researchers. This new field of research will surpass the resolution obtained with conventional fluorescence microscopy, which can distinguish structures in the 200 nm range, to the pioneering use of chromosome conformation captures described by Dekker *et al.* (2002). At the end the ability to detect specific interacting loci will reveal new areas of chromatin and DNA interaction that are unknown at the present time. For example, this interaction has been described as a biological system that juxtaposes separate stretches of DNA that control gene expression. The mediator protein often binds to enhancer sequences and core promoters of genes transcribed in some cell type but not in others. Therefore, greater understanding of these new sets of interactions will brighten our knowledge of the functional genomics. The technologies for this understanding are coming from different manufacturers, but it is crucial that the new cadre of cancer researchers generalize these individual observations so they apply to more than one cell type, both physiological and pathological conditions and that they extrapolate results from few cells to many cells.

2.3. Variations in the Genome

After the human genomic sequencing was finished the sad realization was that one genome was not enough and that a thousand sets of individuals were needed to explain the complexity of not only normal physiologic processes but also human diseases, including cancer. The problem is rooted in the fact that each person contains an average of more than 3 million variants from the reference genome. Adding still more confusion is the fact we do not yet understand what those variants do. Genome-wide association studies have not been enough to determine the biological effects behind these variants, and newly developed techniques are intended to answer this question, for example, using algorithms to predict whether a mutation is likely to change the function of a protein, and sequencing data from larger human populations to reveal which variants can be tolerated by evolution and exist in healthy individuals.

The problem is that ideally the loss of a function due to a mutation could be overridden by the redundancy of the genome, but their importance in clinical diagnosis not yet understood. This does not mean that we should stop searching for the importance of these variants, but only that the answer and application of this knowledge is not immediately at hand. The use of computational tools using evolution to rank variants in noncoding regions is also limited because the comparison is between non-mammal species. Even more complex is the fact that noncoding regions evolve very quickly and thus sequences can be compared only among mammals (Cooper and Shendure, 2011). All these predictive algorithms can tell cancer researchers which variants could be potentially important but still we do not know what relevance there is for specific diseases or types of cancer and thus more data are needed to have enough statistical power to be meaningful.

The next set issue to examine is what the biological effects of these variants are, and this requires laboratory work. For example, making an adequate set of experiments to determine how these variants affect RNA splicing of transcription rates, or the variant proteins that they produce, among other things. An important piece of missing knowledge is whether or not the variations take place in an active part of the genome. To determine this, the ENCODE project (Encyclopedia of DNA Elements) is aiming to map and annotate all functional elements in the genome while the International Cancer Genome Consortium is striving to map all genomic changes in cancer. The International Human Epigenome Consortium and the U.S. National Institutes of Health's Roadmap Epigenomics Mapping Consortium are studying features such as DNA methylation and other modifications across the genome in many types of cell, and so are identifying which regions of the genome might be functional in particular tissues. The next step is to determine if the data obtained in the *in vitro* conditions are applicable to living cells and organisms. These experiments will provide information about averages in a given population but it will not be enough to tell us the meaning of these variations at the individual level. These data are coming but it will require significant efforts by the new cadre of cancer researchers to make them meaningful and useful.

We must acknowledge that several things have been accomplished in the last 5 years, such as the translation of DNA sequencing to the clinic, mainly in oncology for the personalized treatment of cancer, as well as for the identification of genetic abnormalities present in other diseases. Today, there is better software available, which allows managing, securing, and ascertaining results from these large genomic data sets so researchers might make biological sense of all the genomic variations in the context of disease. These advances have allowed physicians to better evaluate the utility of sequencing in the diagnosis, and more accurately target treatment. They have also made insurance companies more willing to pay for these tests. It is the function of the new cadre of cancer researchers to persist in expanding this wave of new knowledge and technology.

2.4. New Frontiers for Immunology and Cancer

The human immune system can be broadly divided into the innate and adaptive immune systems. The latter enables the formation of immune memory through T-cell responses and the production of antibodies by B cells. Although many genetic variants control these processes, the human major histocompatibility complex (MHC) has a predominant role in determining specificity of the adaptive immune response. By contrast, the innate immune system comprises phylogenetically older mechanisms of immediate immune defense with various cellular components, many of which interact with the adaptive immune system. Because of the relevance and the wealth of new developments, I dedicate more space in this chapter to discussing the role of the immune system and cancer.

2.4.1. *The immune system and cancer*

The immune system not only protects the host against development of primary cancers, but also modulates tumor immunogenicity (Chodon *et al.*, 2015; Dunn *et al.*, 2004). The immunobiology of cancer is a dynamic process that can be divided into three phases: elimination, equilibrium, and escape. Elimination represents the classical concept of cancer immunosurveillance (Shankaran *et al.*, 2001), the equilibrium is the period of immune-mediated latency which occurs after incomplete tumor destruction in the elimination phase (MacKie *et al.*, 2003). The escape phase refers to the final outgrowth of tumors that have outstripped immunological restraints of the equilibrium phase (Marincola *et al.*, 2000). A new concept has been added to these three phases, "cancer immunoediting", so-called because of observations that tumor infiltration by lymphocytes is a reflection of a tumor-related immune response. Data from these studies indicate that the presence of tumor-infiltrating lymphocytes (TILs) may be associated with improved clinical outcome in several cancers, including melanoma, colorectal, breast, prostate, renal-cell, esophageal, and ovarian carcinomas (Clemente *et al.*, 1996). A meta-analysis of 10 studies with 1815 ovarian cancer patients confirmed the observation that a lack of intraepithelial lymphocytes (TILs) is significantly associated with a worse survival rate among ovarian cancer patients (Hwang *et al.*, 2012). The major criteria required for the immunological destruction of tumors includes generation of sufficient numbers of effector T cells with high avidity recognition of tumor antigens (TAs) *in vivo*, trafficking and infiltration into the tumor, overcoming inhibitory networks in the tumor microenvironment, and persistence of the antitumor T cells. Meaning that the boosting of an existing immune response against cancer cells has been the basic motif for the development of cancer vaccines, cell-based therapy, immune checkpoint blockade, and oncolytic virus-based therapy.

2.4.2. *The right antigen*

The development of therapeutic approaches has been made possible by defining and understanding the different types of TAs. Differentiation antigens, for example, tyrosinase, are restricted to very defined tissues (Brichard *et al.*, 1993) and multiple epitopes, such as in the human melanoma antigen gp100, by tumor-infiltrating T lymphocytes associated with *in vivo* tumor regression. Another type of TAs are the mutational antigens, which are altered forms of proteins, for example, CDK4 is (Wolfel *et al.*, 1995) a CASP-8 mutation recognized by cytolytic T lymphocytes on a human head and neck carcinoma (Scanlan *et al.*, 1998). Also tumors may amplify antigens, like the case of Her2/neu (Cheever *et al.*, 1995), and the immunity to oncogenic proteins such as P53 (Gnjatic *et al.*, 1998), or splice variant antigens, such as the NY-CO-37/PDZ-45 (Scanlan *et al.*, 1998). Glycolipid antigens and viral antigens such as HPV (Tindle, 1996) have been extremely useful in furthering research in the field. Lastly the cancer-testis (CT) antigens, restricted in expression to the germ line and tumors such as the MAGE (Boon and van der Bruggen, 1996), stimulate cellular and/or humoral responses in cancer patients and give rise to epitopes that are presented on tumor cells (TCs) in the context of the MHC class I or II molecules, thereby stimulating $CD8^+$ or $CD4^+$ T cells, respectively.

The criteria for using these antigens as targets are: (i) limited or no expression in normal tissues, but aberrant expression at high frequencies in tumor; (ii) immunogenicity, and (iii) a role in tumor progression. While none of the current TAs completely meets all of these criteria, the family of CT antigens is the ideal and has a subclass encoded by approximately 140 genes. The criteria for placing antigens in this category are based on several characteristic features (Old, 2001). The CT antigens are known to be restricted in immune privileged sites such as the testes, placenta, and fetal ovary, but not in other normal tissues. Abnormal expression of these germ-line genes in malignant tumors may reflect the activation of a silenced "gametogenic program", which ultimately leads to tumor progression and broad immunogenicity (Simpson *et al.*, 2005).

2.4.3. *Vaccines against cancer*

The development of vaccines against solid and hematological malignancies has been facilitated by the identification and characterization of peptide epitopes from several TAs, along with the relative ease of production of peptides for clinical use. Additional vaccine studies have included the use of long peptides, recombinant proteins, recombinant viral vectors, and dendritic cells (DCs). The main problem has been generating sufficient antitumor immune responses, likely because these TAs are self-proteins and subject to central and peripheral tolerance. Several parallel clinical trials focusing on the prototype antigen NY-ESO-1 were conducted (Old, 2008) to examine this problem. These studies have also raised other relevant questions such as: Which are the appropriate adjuvants? Are monovalent or multiantigen vaccine approaches likely to provide better results? At what disease state is vaccination appropriate? What is the optimal frequency and duration of vaccination? How should antigen-specific immune responses be monitored? How should the induced immune response be sustained?

The most common cancer vaccine strategy is to administer full-length recombinant protein antigens or peptides to cancer patients, most often via the intramuscular, subcutaneous, or intradermal route, together with one or more immunostimulatory adjuvants to promote DC maturation. The rationale behind this approach is that resident DCs or other antigen-presenting cells (APCs) acquire the ability to present the tumor-associated-antigen-derived epitopes while maturing, hence priming a robust tumor associated antigen (TAA)-specific immune response. While short peptides (8–12 amino acids) directly bind to MHC molecules expressed on the surface of APCs, synthetic long peptides (25–30 residues) are taken up, processed, and presented by APCs for eliciting an immune response

(Melief and Van der Burg, 2008). Several reports indicate that therapeutic activity of synthetic long peptides is superior to that of their short counterparts, especially when they include epitopes recognized by both cytotoxic and helper T cells or when conjugated to efficient adjuvants (Melief and Van der Burg, 2008). Both peptide- and DNA-based vaccines have been associated with clinical activity in several cancer types (Galluzzi *et al.*, 2012). For example, the administration of a multipeptide vaccine after single-dose cyclophosphamide was shown to prolong overall survival (OS) in a cohort of renal cell carcinoma patients (Walter *et al.*, 2012). However, the United States Food and Drug Administration (FDA) and European Medicines Agency (EMA) are not approving peptide- or DNA-based anticancer vaccines due to the immune resistance mechanisms that significantly reduce the efficacy of vaccine-induced immune response.

2.4.4. *The role of the DCs*

DC (see Chap. 1) express a number of cytokines and membrane costimulators that drive the T-cell response; DCs also "cross-present" antigens on MHC Class I (Savina *et al.*, 2006). Several forms of DC-based vaccine approaches have been developed, most of which involve the isolation of patient-derived circulating monocytes and their differentiation *ex vivo*, in the presence of agents that promote DC maturation, such as granulocyte macrophage colony-stimulating factor (GM-CSF). These autologous DCs are injected into cancer patients upon exposure to a TA (protein, peptide, mRNAs, recombinant viral vectors encoding TA, TC lysates). The antigen-pulsed DCs are able to prime tumor-targeting immune responses *in vivo* upon administration to patients. An additional strategy is to fuse the tumor to mAbs that selectively bind to endocytosis receptors (e.g., mannose receptor or DEC-205) on the surface of DCs (Tsuji *et al.*, 2011). Only one cellular product containing a significant proportion of probably immature DCs (sipuleucel-T or Provenge) is currently licensed by the U.S. FDA and the EMA for the therapy of asymptomatic or minimally symptomatic metastatic castration-refractory prostate cancer (Kantoff *et al.*, 2010). The safety and efficacy of many other DC-based cellular vaccines are currently being investigated in clinical trials. Three metastatic melanoma trials showed superior survival in patients receiving a therapeutic vaccine consisting of autologous DC loaded with antigens from self-renewing, proliferating, irradiated autologous TCs (DC-TC) compared with patients receiving the autologous irradiated TCs alone (Dillman *et al.*, 2015).

2.4.5. *The neoantigens*

Using next-generation sequencing and epitope prediction has permitted the rapid identification of mutant tumor neoantigens. This has led to efforts in utilizing these mutant tumor neoantigens for personalizing cancer immunotherapies. The infusion of autologous *ex vivo*–expanded TILs can induce objective clinical responses in metastatic melanoma (Dudley *et al.*, 2013). The relationship between pretherapy CD8[+] T cell infiltrates and response to checkpoint blockade in melanoma (Tumeh *et al.*, 2014) produces the programmed cell death-1 (PD-1) blockade by inhibiting adaptive immune resistance. Deep-sequencing technologies permit easy identification of the mutations present within the protein-encoding part of the genome (the exome) of an individual tumor, allowing for prediction of potential neoantigens. Several preclinical and clinical studies have now confirmed the possibility of identifying neoantigens on the basis of cancer exome data (Castle *et al.*, 2012). Epitope presentation of neoantigens by MHC class I molecules may be predicted using previously established algorithms that analyze critical features such as the likelihood of proteasomal processing, transport into the endoplasmic reticulum, and affinity for the relevant MHC class I alleles. Based on these considerations, it becomes of interest to stimulate neoantigen-specific T-cell responses in cancer patients using two possible approaches. The first is to synthesize long peptide vaccines that encode a set of predicted neoantigens. The second approach is to identify and expand preexisting neoantigen-specific T-cell populations to create either bulk neoantigen-specific T-cell products or T-cell-receptor–engineered T cells for adoptive therapy. Adoptive cell transfer (ACT) is an approach that involves: the collection of circulating or TILs T cells (Rosenberg *et al.*, 2008), the activation and modification and/or expansion *ex vivo*; and their reinfusion to patients, usually after lympho-depleting preconditioning chemotherapy. The T-cell receptor (TCR) strategy is based on the understanding that the binding of MHC-antigen complex by

TCR is the main determinant of tumor recognition by T cells. Genes that encode the α and β chains are cloned from tumor-reactive T cells restricted to a particular human leukocyte antigen allele and then introduced into recipient T cells to endow them with the specificity of the donor TCR. Transduced T cells then acquire stable reactivity to the TAAs. The majority of clinical responses are short-lived with ultimate tumor relapse. A major explanation for this suboptimal outcome is the relatively limited long-term survival and effector function due to suppression or exhaustion of infused engineered T cells (see also Robbins *et al.*, 2011; Davila *et al.*, 2014).

2.4.6. *Immune modulation*

One fast-developing area that is the present and future endeavor of the new cadre of cancer researchers is immune modulation that is designed to reinstate an existing anticancer immune response or elicit novel responses as a result of antigen spreading. This reprogramming of the immune system has been achieved through four general strategies: (i) the inhibition of immunosuppressive receptors expressed by activated T lymphocytes, such as cytotoxic T lymphocyte-associated protein 4 (CTLA4) and PD-1; (ii) the inhibition of the principal ligands of these receptors, such as the PD-1 ligand CD274 (PD-L1 or B7-H1); (iii) the activation of costimulatory receptors expressed on the surface of immune effector cells, such as tumor necrosis factor receptor superfamily, member 4 (TNFRSF4 or OX40), TNFRSF9 (CD137 or 4-1BB), and TNFRSF18 (GITR); and (iv) the neutralization of immunosuppressive factors released in the tumor microenvironment, such as transforming growth factor β1.

Checkpoint blockade has been shown to induce robust and durable responses in patients with a variety of solid tumors. Antibody blockade of PD1 and PDL1 has demonstrated enhanced antitumor immunity in mouse models (Blank *et al.*, 2004). PD-L1/B7H-1 inhibits the effector phase of tumor rejection by TCR transgenic CD8$^+$ T cells. A number of checkpoint-blocking mAbs were recently approved by the FDA and other international regulatory agencies for use in humans, namely: the anti-CTLA4 mAb ipilimumab (Yervoy), which was licensed by the FDA for use in individuals with unresectable or metastatic melanoma; the anti-PD-1 mAb pembrolizumab (Keytruda), which received approval by the FDA for the treatment of advanced or unresectable melanoma patients who fail to respond to other therapies; and nivolumab (Opvido), another PD-1-targeting mAb licensed by both the FDA and Japanese Ministry of Health and Welfare for use in humans. Blockades of additional inhibitory receptors are in various phases of clinical development and include LAG3, B7-H3, B7-H4, and TIM3. Emerging evidence suggests that the clinical efficacy of immunomodulatory mAbs (especially checkpoint blockers) may be profoundly influenced by the mutational burden and "neoantigens" specific to the neoplasm (Snyder *et al.*, 2014). The higher neoantigen load leads to recruitment of a diverse repertoire of neoantigen-specific T cells, and checkpoint blockade restores a favorable balance of Teff/Treg ratio, leading to more effective tumor control.

2.4.7. *The checkpoint blockade*

As described earlier, the malignant cells provide neoantigens generated by vast tumor-specific mutation as potential targets for immune destruction by host immunosurveillance. Therefore, checkpoints are a specific subset of negative regulators that normally deliver inhibitory signals to limit effector T-cell response and maintain self-tolerance during antimicrobial immune responses. Malignant cells are able to co-opt immune checkpoint molecules to avoid immune recognition and elimination. It is also known that upon antigen presentation, T cells require the help of costimulatory receptors, such as CD28, to be fully activated. In addition, a wide range of coinhibitory receptors are also expressed on T cells to avoid collateral tissue damage caused by an over-activated immune response. Together, these molecules form a complex network of signals to balance T-cell activation, tolerance, and immunopathology. CD28 is the first and most pronounced costimulatory receptor that is expressed on T cells and this molecule participates in the classical two-signal model of T-cell activation as has been discussed scholarly by Linsley and Ledbetter (1993). During the normal immune response, CD28 colocalizes with the TCR in response to presentation of peptide-MHC complexes by APCs to form the immunological synapse where CD28-TCR acts as the secondary stimulus to prompt TCR signaling (Linsley and Ledbetter, 1993). However, two ligands of CD28, CD80 (B7-1), and CD86 (B7-2), are also ligands of CTLA-4 which, on the contrary, is a well-known coinhibitory receptor that suppresses T-cell function during the early phase of activation (Teft *et al.*, 2006). CTLA-4 and CD28 are located next to each other on chromosome 2q33 (Naluai *et al.*, 2000). The sharing of ligands between CD28 and CTLA-4 underlines the major role of CTLA-4 in preventing CD28-induced stimulatory signals. The balance of CTLA-4 and CD28 is under accurate spatiotemporal regulation during the immune response. Unlike CD28, CTLA-4 expression is mainly cytoplasmic and absent on the surface of naïve resting T cells to allow initiation of immune activation upon antigen recognition. CTLA-4 is then mobilized to the T-cell surface and reorients to the central supramolecular activation cluster (c-SMAC) of immunological synapse where it competitively sequesters CD80 and CD86 from CD28, thereby suppressing CD28-dependent costimulation (Teft *et al.*, 2006; Egen and Allison, 2002). CTLA-4 has much greater affinity for both ligands compared with CD28, suggesting inhibitory signals have a dominant role over stimulatory signals (Collins *et al.*, 2002).

CTLA-4 resides primarily in the cytoplasm and the level of expression of CTLA-4 on the T-cell surface is mostly limited and insufficient to abrogate T-cell activation (Freeman *et al.*, 2000). CTLA-4 signaling also increases T-cell motility and reduces the TCR stop signal required for the maintenance of a stable immunological synapse between T cells and APCs, consequently limiting the requisite signals for T-cell activation (Schneider *et al.*, 2006). Additionally, CTLA-4 can capture shared CD28-ligands from APCs and degrade them within CTLA-4 expressing T cells via a process called "trans-endocytosis" (Qureshi *et al.*, 2011). CTLA-4 initiates reverse signals through CD80/CD86 to induce the expression of indoleamine 2,3-dioxygenase (IDO) in APCs, leading to IDO-induced localized tryptophan depletion that suppresses conventional T cells while promoting the function of regulatory T cells (Tregs) (Munn *et al.*, 2014).

PD-1 is the second checkpoint molecule that has been clinically targeted. PD-1 mainly acts as a brake on effector T cells and reduces immune responses in the tissue microenvironment (Rudd *et al.*, 2009). The phenotype of PD-1 knockout mice is characterized by late-onset, organ-specific autoimmunity that sharply contrasts with the rapid-onset fatal multiorgan inflammation of the CTLA-4 knockout mice (Nishimura *et al.*, 2001; Intlekofer and Thompson, 2013). PD-1 is expressed on the cell surface of different immune cells including activated T cells, B cells, and natural killer T cells. PD-1 has two ligands, PD-L1 (B7-H1) and PD-L2 (B7-DC). PD-L1 is widely expressed by hematopoietic and non-hematopoietic cells whereas PD-L2 is inducibly expressed on DCs (Keir *et al.*, 2008).

Upon ligand binding, CTLA-4 reorients to the immunological synapse and excludes CD28 from the central c-SMAC. Phosphorylation of CTLA-4 at YKVM motif recruits PP2A and SHP-2, which further dephosphorylate early TCR-signaling components. CTLA-4 does not disturb PI3K-mediated antiapoptosis effect, but directly targets Akt. PD-1, in contrast, can only drive CD28 away, but not exclude it from the c-SMAC. Activated PD-1 recruits SHP-1 and SHP-2 to its phosphorylated immunoreceptor tyrosine-based inhibitory motif (ITIM) and immunoreceptor tyrosine-based switch motif (ITSM). The ligation dephosphorylates the early TCR signaling in a similar way with CTLA-4. However, PD-1 can directly block TCR/CD28-mediated activation of PI3K. Together, CTLA-4 and PD-1 contribute to the coinhibitory effect toward T cells, resulting in impairment of T-cell activation, subset differentiation, effector function, and survival. Similar to CTLA-4, PD-1 engagement can restrain the duration of migration arrest of T cells on APCs (Honda *et al.*, 2014). PD-L1 is an additional ligand of CD80 and therefore also can competitively inhibit the costimulatory signal provided by CD28-CD80 ligation (Butte *et al.*, 2007).

2.4.8. Oncolytic viruses

In addition to the checkpoint blockade the use of oncolytic viruses has also emerged as an immunotherapy tool. Oncolytic viruses are nonpathogenic viral strains that specifically infect cancer cells, triggering their demise. The antineoplastic potential of oncolytic viruses can be innate via a cytopathic effect, or these viruses can mediate an oncolytic activity because of gene products that are potentially lethal for the host cell, irrespective of the host's capacity to massively replicate and cause a cytopathic effect. Increasingly preclinical and clinical evidence indicate that the therapeutic activity of oncolytic viruses is also related to their ability to elicit tumor-targeting immune responses as they reprogram the inflammatory tumor microenvironment to be more immunogenic and also promote the release of TAs in this immunostimulatory environment, leading to efficient cross presentation. These viruses can be genetically engineered to endow them with additional attributes such as antagonist of chemokine–chemokine ligand interaction (Gil *et al.*, 2014). CXCL12/CXCR4 blockade by oncolytic virotherapy inhibits ovarian cancer growth by decreasing immunosuppression and targeting cancer-initiating cells. The results of immunotherapy with talimogene laherparepvec (T-VEC), a modified form of herpes simplex virus type-1, which provides direct tumor lysis, that simultaneously produce GM-CSF were reported by Andtbacka *et al.* (2015). T-VEC improves durable response rate in patients with advanced melanoma. Researchers randomized 436 patients with aggressive, inoperable malignant melanoma to receive either an injection of the viral therapy T-VEC, or a control immunotherapy in a phase III trial. Of the group given T-VEC, 16.3% showed a durable treatment response of more than 6 months. Some patients had a response extending past 3 years.

Combination therapy with adoptive T-cell therapy and oncolytic viral delivery may have beneficial synergistic effect. In mouse models, it has been shown that antigen-nonspecific T cells loaded with oncolytic vesicular stomatitis virus efficiently delivered the virus to metastatic lymph nodes, leading to tumor clearance associated with antitumor immune priming (Qiao *et al.*, 2008a). A combination of antigen-nonspecific adoptive T-cell therapy, oncolytic virotherapy, and immunotherapy has been shown to purge metastases in lymphoid organs. The loading of antigen-specific T cells with vesicular stomatitis virus enhanced the delivery of the virus to lung tumors (Qiao *et al.*, 2008b). Loading oncolytic vesicular stomatitis virus onto antigen-specific T cells enhances the efficacy of adoptive T-cell therapy of tumors and the associated pro-inflammatory tumor microenvironment enhanced antigen-specific T-cell proliferation and survival within the tumor.

The new cadre of cancer researchers has at their disposal an arsenal of new tools that could be personalized to treat cancer with immunotherapy, a technique that could one day become the standard of care for cancer patients. From all the information discussed earlier, it is clear that immunotherapy is expected to not only mediate tumor destruction and drive local inflammation in the tumor microenvironment, but also to trigger coordinated induction of multiple counter-regulatory and suppressive pathways like IDO, TGFβ, PD-L1, and Tregs. Therefore, a concomitant blockade of these suppressive pathways mediated by immune suppression at the time of vaccination or T-cell transfer will allow inflammation-induced transformation of the tumor milieu to be a powerful target of cancer treatment.

Based on the promising results of blockade of the PD-1/PD-L1 pathway, it is important to consider opportunities for combination therapies. These include dual checkpoint blockade, for example, the combination of CTLA-4 and PD-1 blockade. Ipilimumab removes a physiological brake on T cells during activation, whereas anti-PD-1 removes a brake on activation during T-cell effector function. This combination may also overcome resistance to CTLA-4 blockade mediated by tumor PD-L1 expression or resistance to PD-1 blockade mediated by T-cell downregulation through the coexpression of CTLA-4 (Larkin *et al.*, 2015). Another potential checkpoint combination therapy is blockade of PD-1 and LAG-3, an approach that has demonstrated excellent results in preclinical models of ovarian cancer and melanoma (Huang *et al.*, 2015). LAG3 and PD1 coinhibitory molecules collaborate to limit CD8+ T-cell signaling and dampen antitumor immunity in a murine ovarian cancer model.

2.4.9. *Side effects of the immune checkpoint blockade*

Monoclonal antibodies targeted against the immune checkpoint molecules CTLA-4 and PD-1 are used for the treatment of metastatic melanoma and advanced/refractory non–small-cell lung cancers (NSCLC). Beyond melanoma and NSCLC, immune checkpoint blockaders (ICBs) are showing promising responses across many different cancer subtypes, including small-cell lung cancer (15% overall recurrence rate (ORR)) (Antonia *et al.*, 2016), renal cell carcinoma (25% ORR) (Motzer *et al.*, 2015), urothelial cancer (25% ORR) (Plimack *et al.*, 2015), head and neck squamous cell carcinoma (HNSCC) (12%–25% ORR), hepatocellular carcinoma (20% ORR) (El-Khoueiry *et al.*, 2015), ovarian cancer (15% ORR) (Hamanishi *et al.*, 2015), and triple negative breast cancer (20% ORR) (Emens *et al.*, 2015). However, despite their encouraging efficacy profiles, these immune-targeted agents also generate immune-related adverse events (irAEs). This new family of dysimmune toxicities remains largely unknown to the broad oncology community and although severe irAEs remain rare (~10% of cases under monotherapy), they can become life-threatening if not anticipated and managed appropriately.

2.5. Synthetic Biology

Synthetic biology has been defined in simple terms as a more complex form of genetic engineering, or a way to design biological products using engineering principles and a modular approach. At the present time, there are those who advocate free-for-all tools, such as open-source repositories, and the groups that seek to protect the fruits of research using legal instruments such as patents. Some of those representing the field of synthetic biology believe that the patent-heavy intellectual-property model of biotechnology is hopelessly broken (Nelson, 2014). They are instead relying on freely available software and biological parts that could be combined in innovative ways to create individualized cancer treatments — without the need for massive upfront investments or a thicket of protective patents. "The other view is more akin to what happens at big pharmaceutical companies, where revenues from blockbuster drugs fund massive research initiatives behind locked doors. For such businesses, the pursuit of new drugs and other medical advances depends heavily on protecting discoveries through patents and restrictive licensing agreements" (Nelson, 2014). As Nelson indicates, how synthetic biologists resolve the conflict between open source and patent protection could determine whether or not the field delivers on its ambitious goal of transforming medicine, agriculture, energy, environmental remediation, and other industries through precision engineering.

According to Nelson (2014) "the field of systemic biology is an amalgam of diverse influences, each with different cultures of intellectual property. On one side sit software design and engineering, which introduced the idea of encoding desired functions in pieces of DNA and joining together a standardized set of biological widgets, much like bricks or Lego pieces. Software engineers also brought with them the philosophy of sharing their work using open, public registries or only lightly restrictive licensing agreements, such as copyrights. On the other side sit molecular biology and biotechnology, which supplied know-how about messy and unpredictable biological systems. They also brought the practice of patenting genes, molecules and technical processes". Legally, therefore, synthetic-biology sequences and techniques can be patented, at least in the United States. But the morals and ethics of doing so are vigorously debated by researchers, companies, lawyers, and bioethicists.

A more radical problem in the emerging field of synthetic biology according to Collins (2014) is that there are few biologists. To be fair, the engineering slant of synthetic biology has brought impressive accomplishments. These include whole-cell biosensors; cells that synthesize antimalaria drugs; and bacterial viruses designed to disperse dangerous, tenacious biofilms. However, these engineer designs are based on model systems as black boxes, abstractly linking inputs and outputs. They can often control a system with only a limited understanding of it. But synthetic-biology projects are frequently thwarted when engineering runs up against the complexity of biology. The reason of these concerns is that in real life, organisms comprise a hierarchy of systems from the subcellular level to the whole body, and many models have been developed across this organismal hierarchy to address-specific questions across biology and medicine. One of them is the organoid model that provides a unique opportunity to incorporate moderate system complexity while still affording many tools for probing structure and function (Lancaster and Knoblich, 2014). When compared to tissue explants, organoid systems can mimic similar cell–cell and cell–matrix interactions while maintaining the ability for long-term cultures thanks to maintained signaling cues important for survival. While a wide variety of organoids have been generated, most organoid models only represent single or partial components of a tissue, and it is often difficult to control the cell type, organization, and cell–cell or cell–matrix interactions within these systems. Bioengineers have long aspired to deconstruct biological systems then manipulate or reconstruct the system in a controlled manner. Bioengineering approaches have enabled us to steer cell behavior and cell organization, which are fundamental processes in organoid formation, and improved systems are on the horizon. Organoids have been generated from both pluripotent stem cells (PSCs) and adult stem cells (ASCs) by mimicking the biochemical and physical cues of tissue development and homeostasis (Lancaster and Knoblich, 2014). In very simplified view, the development of the human body is a precisely controlled process of step-wise differentiation from the zygote and the subsequent self-organization of the cells generated in this process. The process can be partially reproduced when PSCs form a teratoma containing a variety of semiorganized tissues following uncontrolled differentiation and self-organization (Przyborski, 2005). Similarly, this process can be controlled *in vitro* with PSCs induced to differentiate down specific lineages. If provided the proper 3D scaffold and biochemical factors, differentiated cells from PSCs will self-organize to form tissue-specific organoids including the optic cup (Eiraku *et al.*, 2008), brain (Lancaster *et al.*, 2013), intestine (Spence *et al.*, 2014), liver (Takebe *et al.*, 2013), and kidney (Takasato *et al.*, 2014). Additionally, the homeostasis of many tissues *in vivo* is maintained by tissue-specific ASCs through self-renewal and differentiation, followed by self-organization of the stem cells and their progeny. This process can also be reproduced *in vitro* under specific culture conditions to control self-renewal and differentiation, resulting in self-organized tissue organoids including intestine (Sato *et al.*, 2009), stomach (Barker *et al.*, 2010), liver and pancreas (Huch *et al.*, 2013), and prostate (Karthaus *et al.*, 2014). During the establishment of organoid model systems, several bioengineering approaches that were developed in other fields, including stem cell niche engineering and tissue engineering, became available to steer the behavior and organization of organoids. A major aim of engineering organoid systems is to improve system utility in downstream applications. Thus, effort is required to create better proxies for *in vivo* tissues and organs and improve organoid system modularity to accommodate high-throughput formats or even multitissue organoid compatibility in larger multiplexed systems (e.g., human-on-a-chip) (Lancaster and Knoblich, 2014).

2.6. New Models in Cancer Research

Despite the many advances in our understanding of cells and tissues produced by the research endeavor, many diseases remain incurable. As Alberts (2010) indicates "there is a disparity between the enormous amount now known about the chemistry and molecular biology of cells and our ability to intervene in human disease may seem incongruous to the public, but it is not at all surprising to the scientists involved: As we have learned more about how cells work, we have been surprised to discover how enormously sophisticated and complex are the processes that produce a human being".

Undoubtedly, the availability of the human genome sequence has transformed biomedical research over the past decade; now an equivalent map for the human proteome with direct measurements of proteins and peptides has been published using high-resolution Fourier-transform mass spectrometry. Analysis of the complete human genome sequence has thus far led to the identification of approximately 20,687 protein-coding genes, although the annotation still continues to be refined. The generated tandem mass spectra corresponds to proteins encoded by 17,294 genes accounting for approximately 84% of the annotated protein-coding genes in the human genome — to our knowledge the largest coverage of the human proteome reported so far (Kim *et al.*, 2014). Each of these proteins in their different network of activities is a potential target for drug intervention. The screening of new drugs has been based on the use of the NCI-60, a panel of 60 human cancer cell lines implemented by the NCI for the last 25 years. As useful as this has been in the testing of more than 100,000 compounds, in order that we make medical discoveries that create a real difference in the lives of a great number of people, we must be sure that what we discover in the test tube and the *in vitro* system of cell culture is demonstrated in the animal system. For this reason, the NCI-60 will be replaced by a repository of cancer models that are derived from fresh patient samples and tagged with details about their clinical pasts. This decision was based in the need for cancer models with a closer link to the patients they are intended to help (Ledford, 2016). According to Ledford (2016) "the new model is based on the use of hundreds of patient-derived xenografts (PDXs) by implanting small pieces of human tumors in mice — an environment that better mimics the human body. The tumors can then be harvested and reimplanted in other mice, allowing researchers to study a given tumor in multiple animals". Although this process is already used by many laboratories around the world, only time will determine the real impact of this method in choosing the right drug alone or in combination. In addition, the most reliable animal studies are those that use randomization to eliminate systematic differences between treatment groups; induce the condition under investigation without knowledge of whether or not the animal will get the drug of interest; and assess the outcome in a blinded fashion (Macleod, 2011). Studies that do not report these measures are much more likely to overstate the efficacy of interventions and the other problem is that many of the studies reported are underpowered. This also is a test that the new cadre of cancer researchers must evaluate.

The main concern in the scientific community is that all the data obtained in experimental animals, mainly mice and rats, cannot be directly extrapolated to humans, and in fact we can cure tumors in mice and rats but not in humans. The use of nonhuman primates has traditionally been the option to make studies nearer to the human situation, however, the number of monkeys used in cancer research and medical research in general has decreased significantly due to the tangle of regulatory hurdles, financial constraints, and bioethical opposition According to Cyranoski (2016) China has emerged as a potential center for performing research in nonprimates, whereas in the National Cancer Institute (NCI) and other educational institutions they are closing or significantly reducing the facilities for studying monkeys. As it is analyzed by Cyranoski (2016) the cost is a factor; for example, it costs $6000 to buy a monkey in the United States, and $20 per day to keep it, whereas the corresponding figures in China are $1000 to buy and $5 per day. This in itself is an incentive to move research enterprise to China, but in addition nonhuman primate research centers have been able to construct first chimeric monkeys using embryonic stem cells, an advance that could make the production of genetically modified animals even easier" and in addition "the rapid spread of CRISPR-Cas9 and TALEN gene-editing tools is likely to accelerate demand for monkey research". All of these improvements are "turning the genetic modification of monkeys from a laborious and expensive task into a relatively quick, straightforward one" (Cyranoski, 2016).

For cancer researchers it would not be easy to move research enterprises to China and the cost and infrastructure will be an important component of the final equation. Thus the new cadre of cancer researchers must contemplate not only the use of nonprimates in cancer research but also the feasibility of funding research outside of the United States or Europe.

2.7. The Stem Cell

Induced pluripotent stem (iPS) cells, which are created by "reprogramming" adult cells into a stem-cell-like state, offer a way to skirt the thorny ethical issues raised by extracting cells from human embryos. The idea behind this is that the adult cells taken from a patient could be used to create stem cells that would, in turn, generate perfectly matched specialized tissues (Baker, 2013). According to Buczacki *et al.* (2013) a good example of stem cells is found in the crypt–villus units of the small intestine, their self-renewal takes 4–5 days. There are two stem-cell populations (one rapidly cycling, the other quiescent). ASCs must be capable of self-renewal and persistence over the lifetime of an animal, and also be multipotent — have the capacity to generate all cell types of a tissue. After their formation from stem cells at crypt bottoms, young epithelial cells proliferate intensely for 2 days before differentiating and exiting the crypts. The cells then moves on to the villus flanks to process nutrients and finally die at the villus tips on days 4-5. Paneth cells also derive from stem cells, but persist for 1-2 months at the crypt base — in which they are thought to maintain stem-cell renewal and provide protection against bacteria.

iPS cells, created by turning back the developmental clock on adult tissues, and embryonary stem cells (ES) display similar gene expression patterns, and both can produce any of the various tissues in the human body. However, iPS cells may not be suitable substitutes for ES cells in modeling or treating disease due that the patterns of DNA methylation, a type of epigenetic change, cross the genomes of 15 cell lines including four human ES cell lines, five iPS cell lines, and the tissues from which they came, as well as differentiated cells made from both kinds of stem cells, resembled those in the adult tissues from which the iPS cells had been derived. This could constrain the types of tissues that the cells are capable of forming. In the human cells, the epigenetic marks lingered even after the iPS cells had been coaxed to form new tissues. Regardless of their epigenetic differences, it is possible that neither iPS cells nor ES cells will turn out to be perfect models of tissues in the body, as both cell types seemed to harbor genomic abnormalities (Laurent *et al.*, 2011). ES cells tended to contain duplicated chunks of DNA linked to genes associated with self-renewal, whereas iPS cells incorporated extra cancer-causing genes and fewer tumor suppressor genes.

2.7.1. *Stem cells and cancer*

Evidence suggests the presence of a small population of cancer stem cells (CSCs), or tumor-initiating stem-like cells, present in the tumor, capable of self-renewal and generation of differentiated progeny. The presence of these CSCs can be attributed to the failure of cancer treatments as these cells are believed to exhibit therapy resistance. As a result, increasing attention has been given to CSC research to resolve the therapeutic problems related to cancer. Progress in this field of research has led to the development of novel strategies to treat several malignancies and has become a hot topic of discussion. The concept that cancers result from abnormal changes in normal developmental processes is very old. Research has shown that cancer cells are not all the same. The majorities of cells in tumors are nontumorigenic and are marked by limited self-renewal ability. Only a small subpopulation of cancer cells has the ability to self-renew and initiate tumors. These cells are referred to as CSCs, or tumor-initiating cells. This subset of cancer cells displays two hallmark properties of stem cells: self-renewal ability and the capacity to differentiate. Although less than 1% of the overall cancer cells have the ability to proliferate extensively and form tumors, they are the major reason for the relapse of tumors, resistance to therapies, and metastasis (Wu *et al.*, 2012; Clarke, 2005). These CSCs may give rise to two identical daughter CSCs by undergoing a symmetrical division, or they may undergo an asymmetrical division to give rise to one daughter CSC and one differentiated progenitor cell, thus increasing the number of CSCs accompanied by growth and expansion of tumor (Clarke *et al.*, 2006).

The existence of CSCs was first proposed by Dick (2008) in hematological malignancies. Since then more and more evidence has emerged to support the existence of CSCs. Recent studies have shown the existence of CSCs in tumors of the brain, breast, prostate, pancreas, hepatobiliary, and colorectal cancer (Han *et al.*, 2013). Several different theories have been postulated regarding the origin of CSCs. One theory believes that normal stem/progenitor cells give rise to CSCs by obtaining the ability to generate tumors after encountering a special genetic mutation or environmental alteration (Kumar *et al.*, 2014). These mutations may occur as a result of genomic instability or induced plasticity through oncogenes. The accumulation of such mutations can enable the cells to acquire the ability of self-renewal and tumorigenicity. Another mechanism believed to generate CSCs, the epithelial–mesenchymal transition (EMT), is characterized by a series of steps, wherein initially fibroblast-like motile cells are formed through transformation of epithelial cells, which eventually acquire the ability to invade, migrate, and disseminate (Han *et al.*, 2013). Aside from being similar to stem cells with respect to self-renewal and production of differentiated progeny, CSCs are also similar to SCs in expression of specific surface markers and in utilization of common signaling pathways. However, CSCs can form tumors when transplanted into animals; normal stem cells, on the other hand, cannot. Tumorigenic activity is thus a major significant difference among the two (Li *et al.*, 2009). Tumors grown from tumorigenic cancer cells can be serially passaged and shown to have a mixed population of both tumorigenic and nontumorigenic cells, thus maintaining the phenotypic heterogeneity of the parent tumor (Hamaï *et al.*, 2014). Aside from the cellular heterogeneity, CSCs ensure survival under genotoxic stress and therapeutic toxicity. This resistance is attributed to removal of chemotherapeutic agents by drug efflux pumps, which are involved in sequestering the drugs and hindering them from reaching transformed cells. Furthermore, CSCs have enhanced DNA damage repair machinery (Han *et al.*, 2013). Cumulatively, these events result in sustenance of transformed cells. Thus, CSCs properly explain the heterogeneity of tumors, its mechanism of relapse and metastasis, and also the poor outcome of current therapies. Therefore the ultimate goal of CSC research should be to focus on isolating and destroying these CSCs, yet another task for the new cadre of cancer researchers.

The main markers used for isolation and identification of CSCs include CD133, CD24, hyaluronic acid receptor, CD44, transcription factors such as OCT-4, SOX-2, and drug efflux pumps such as ATP-binding cassette (ABC) drug transporters and multidrug resistance transporter 1 (MDR1). The CD24 and CD133 markers are widely used to identify CSCs in breast and colorectal cancer (Medema, 2013). Fluorescence-activated cell sorting (FACS), flow cytometry, immunofluorescent staining, and polymerase chain reactions are all widely used to isolate and characterize CSCs (Han *et al.*, 2013). The concomitant study of these molecular markers of stem cells, however, is mandatory so as to characterize them completely.

When cells harvested from tumor specimens are grown in a serum-free media supplemented with epidermal growth factor (EGF) and basic fibroblast growth factor (bFGF), CSCs, as well as normal stem cells, either form spheres or grow into colonies. The CSC population in small-cell lung cancer (SCLC) cell lines H446 cells was suggested by Qiu *et al.* (2012). They demonstrated that the *in vitro* clonogenic and *in vivo* tumorigenic potentials, as well as the drug resistance, had increased in H446 cell lines when they were grown in a defined serum-free medium (Qiu *et al.*, 2012). However, a major limitation associated with this assay is that it does not detect quiescent stem cells. Further, it does not reflect the actual readout of *in vivo* stem cell frequency (Pastrana *et al.*, 2011).

Despite its merits, *in vitro* assays have several limitations and therefore the results of *in vitro* assays must be confirmed with *in vivo* assay. Thus, assays that emphasize the self-renewal and tumor propagation *in vivo* need to be standardized. One assay that fulfills both these criteria is the serial transplantation assay in animal models. In this assay, TCs are transplanted into an immunocompromised mouse and the mouse is monitored at various time points for tumor growth. The xenograft tumors must then be isolated from this mouse and transplanted into another immunocompromised mouse and this mouse should be monitored for self-renewal and tumor initiation. Despite its efficiency, one disadvantage associated with this assay is that they are confounded by variables such as homing efficiency of donor cells and thus are better suited to studies of wild-type cells and not mutant cells (Purton and Scadden, 2007).

Since normal stem cells and CSCs share a common feature of self-renewal, it is believed that these cells may also share the same signaling pathways. The major pathways involved in maintenance and plasticity of CSCs include the Wnt (Klaus and Birchmeir, 2008), Notch, and hedgehog pathways (Hebrok, 2003). In addition to facilitating the crosstalk between the receptor tyrosine kinase (RTK) pathway and interleukin-6 (IL-6), Janus kinase 1 (JAK1), a signal transducer and activator of transcription 3 (STAT-3) signaling, plays a central role in regulating CSC plasticity in solid tumors (Hamaï *et al.*, 2014).

Current anticancer therapies inhibit cancer cell growth, cause cancer cells to die, or a combination of both. Although initial treatments appear to be successful, a relapse generally occurs at a later date. This relapse and resistance to therapy occurs because most traditional and mainstream therapies do not target CSCs. Therefore, it is essential to target these CSCs to prevent tumor relapse and provide an efficient and less toxic treatment for cancer therapy. Various strategies may be employed to target CSCs, these include drugs such as cyclopamine and GDC-0449 (Vismodegib) that inhibit the signaling molecule smoothened (Smo) in the hedgehog pathway. These drugs are generally given in combination with arsenic trioxide (As_2O_3) to increase the efficiency of the treatment (Singh *et al.*, 2011).

Another way to go after stem cells is by targeting miRNAs using antisense oligonucleotide (ASO) inhibition. Nozawa *et al.* (2006) showed that siRNA could downregulate EGFR and inhibit HNSCC; they also showed that this increased the sensitivity of HNSCC to cisplatin, 5-FU, and docetaxel. Apoptosis can be induced in CSCs by p53-based drug therapy or by targeting survivin.

P53 protein in its wild type is responsible for tumor suppression; however, mutation in p53 leads to a gain of oncogenic function. Such mutated p53 can be targeted by several drugs so as to restore the wild-type p53. One example is Phikan083 (Boeckler *et al.*, 2008). Survivin, on the other hand, is an inhibitor of apoptosis protein which is overexpressed in various cancers; as a result, cancer cells are resistant to apoptosis. Targeting survivin for cancer intervention is made possible through the use of ASOs, which sensitize these cells to chemotherapy (Sharma *et al.*, 2005).

2.7.2. *Stem cells and therapeutic resistance*

CSCs are not only responsible for initiation, development, and metastasis of tumors, they also attribute to therapeutic resistance. Therefore, current treatments that do not target the CSCs fail to completely cure cancer. In order to control the aggressiveness of cancer, novel therapeutic agents should target both CSCs and important molecules in the signaling pathways. Such a combination will likely yield dramatic results. Furthermore, since CSCs share many properties with normal stem cells, targeting CSCs may affect normal stem cells as well. Thus, high-precision therapies selectively targeting CSCs while sparing normal stem cells need to be devised.

2.7.3. *Regulation of stemness*

A basic area that still needs attention is understanding the physiological process of regulation of normal stem cells in the body. There are studies pointing to the importance of intrinsic and extrinsic factors in controlling stem-cell divisions (Tomasetti and Vogelstein, 2015; Wu *et al.*, 2016) and therefore linking them to cancer risk. Among these factors is diet, mainly a high-fat diet that could control stem cell proliferation in the intestines of mice, allowing an association between obesity and cancer risk (Beyaz *et al.*, 2016). Although this chapter is not the place to discuss all the possible physiological factors that could control the proliferation of stem cells, it is important to call attention to this specific topic of research.

References

Alberts, B. Model organisms and human health. *Science*, **330**: 1724, 2010.

Andtbacka, R.H., Kaufman, H.L., Collichio, F. Talimogene laherparepvec improves durable response rate in patients with advanced melanoma. *J. Clin. Oncol.* **33**: 2780–2788, 2015.

Antonia, S.J., López-Martin, J.A., Bendell, J., Ott, P.A., Taylor, M., Eder, J.P., Jäger, D., Pietanza, M.C., Le, D.T., de Braud, F., Morse, M.A., Ascierto, P.A., Horn, L., Amin, A., Pillai, RN., Evans, J., Chau, I., Bono, P., Atmaca, A., Sharma, P., Harbison, C.T., Lin, C.S., Christensen, O., Calvo, E. Nivolumab alone and nivolumab plus ipilimumab in recurrent small-cell lung cancer (CheckMate 032): A multicentre, open-label, phase 1/2 trial. *Lancet Oncol.* **17**: 883–895, 2016.

Baker, M. Court lifts cloud over embryonic stem cells. *Nature*, **491**: 282, 2013.

Barker, N., Huch, M., Kujala, P., Van de Wetering, M., Snippert, H.J., Van Es, J.H., Sato, T., Stange, D.E., Begthel, H., Van den Born, M., Danenberg, E., van den Brink, S., Korving, J., Abo, A., Peters, P.J., Wright, N., Poulsom, R., Clevers, H.Lgr5(+ve) stem cells drive self-renewal in the stomach and build long-lived gastric units *in vitro. Cell Stem Cell.* **6**: 25–36, 2010.

Barton, M., Santucci-Pereira, J., Russo, J. Molecular pathways involved in pregnancy-induced prevention against breast cancer. *Front Endocrinol (Lausanne).* **5**: 213–219, 2014.

Beyaz, S., Mana, M.D., Roper, J., Kedrin, D., Saadatpour, A., Hong, S.J., Bauer-Rowe, K.E., Xifaras, M.E., Akkad, A., Arias, E., Pinello, L., Katz, Y., Shinagare, S., Abu-Remaileh, M., Mihaylova. M.M., Lamming, D.W., Dogum, R., Guo, G., Bell, G.W., Selig, M., Nielsen, G.P., Gupta, N., Ferrone, C.R., Deshpande, V., Yuan, G.C., Orkin, S.H., Sabatini, D.M., Yilmaz, Ö.H. High-fat diet enhances stemness and tumorigenicity of intestinal progenitors. *Nature*, **531**: 53–58, 2016.

Blank, C., Brown, I., Peterson, A.C. PD-L1/B7H-1 inhibits the effector phase of tumor rejection by T cell receptor (TCR) transgenic CD8+ T cells. *Cancer Res.* **64**: 1140–1145, 2004.

Boeckler, F.M., Joerger, A.C., Jaggi, G., Rutherford, T.J., Veprintsev, D.B., Fersht, A.R. Targeted rescue of a destabilised mutant of p53 by an in silico screened drug. *Proc. Natl. Acad. Sci. USA.* **105**: 10360–10365, 2008.

Boon, T., Van der Bruggen, P. Human tumor antigens recognized by T lymphocytes. *J. Exp. Med.* **183**: 725–729, 1996.

Brichard, V., Van Pel, A., Wolfel, T. The tyrosinase gene codes for an antigen recognized by autologous cytolytic T lymphocytes on HLA-A2 melanomas. *J. Exp. Med.* **178**: 489–495, 1993.

Buczacki, S.J., Zecchini, H.I., Nicholson, A.M., Russell, R., Vermeulen, L., Kemp, R., Winton, D.J. Intestinal label-retaining cells are secretory precursors expressing Lgr5. *Nature*, 495: 65–69, 2013.

Butte, M.J., Keir, M.E., Phamduy, T.B., Sharpe, A.H., Freeman G.J. Programmed death-1 ligand 1 interacts specifically with the B7-1 costimulatory molecule to inhibit T cell responses. *Immunity*, 27: 111–122, 2007.

Carter, D., Chakalova, l., Osborne, C.S., Dai, Y.-F., Fraser, P. Long-range chromatin regulatory interactions *in vivo*. *Nature Genet.* **32**: 623–626, 2002.

Castle, J.C., Kreiter, S., Diekmann, J. Exploiting the mutanome for tumor vaccination. *Cancer Res.* **72**: 1081–1091, 2012.

Cheever, M.A., Disis, M.L., Bernhard, H. Immunity to oncogenic proteins. *Immunol. Rev.* **145**: 33–59, 1995.

Chodon, T., Koya, R.C., Odunsi, K. Active immunotherapy of cancer molecular and cellular. *Immunology*, **44**: 817–836, 2015.

Clarke, M.F. Self-renewal and solid-tumor stem cells. *Biol. Blood Marrow Transplant.* **11**: 14–16, 2005.

Clarke, M.F., Dick, J.E., Dirks, P.B., Eaves, C.J., Jamieson, C.H., Jones, D.L., Visvader, J., Weissman, I.L., Wahlet, G.M. Cancer stem cells — perspectives on current status and future directions: AACR workshop on cancer stem cells. *Cancer Res.* **66**: 9339–9344, 2006.

Clemente, C.G., Mihm, M.C. Jr., Bufalino, R. Prognostic value of tumor infiltrating lymphocytes in the vertical growth phase of primary cutaneous melanoma. *Cancer*, **77**: 1303–1310, 1996.

Collins, A.V., Brodie, D.W., Gilbert, R.J., Iaboni, A., Manso-Sancho, R., Walse, B., Stuart, D., van der Merwe, P.A. The interaction properties of costimulatory molecules revisited. *Immunity*, **17**: 201–210, 2002.

Collins, J.J. How best to build a cell. *Nature*, **509**: 155–156, 2014.

Cooper, G.M., Shendure, J. Needles in stacks of needles: Finding disease-causal variants in a wealth of genomic data. *Nat. Rev. Genet.* 12: 628–640, 2011.

Cyranoski, D. Monkey kingdom. *Nature.* 532: 300–302, 2016.

Davila, M.L., Riviere, I., Wang, X. Efficacy and toxicity management of 19-28z CAR T cell therapy in B cell acute lymphoblastic leukemia. *Sci. Transl. Med.* 6: 224–225, 2014.

Dekker, J., Rippe, K., Dekker, M., Kleckner, N. Capturing chromosome conformation. *Science,* **295**: 1306–1311, 2002.

Dick, J.E. Stem cell concepts renew cancer research. *Blood.* **112**: 4793–4807, 2008.

Dillman, R.O., McClay, E.F., Barth, N.M. Dendritic versus tumor cell presentation of autologous tumor antigens for active specific immunotherapy in metastatic melanoma: Impact on long-term survival by extent of disease at the time of treatment. *Cancer Biother. Radiopharm.* 30: 187–194, 2015.

Dudley, M.E., Gross, C.A., Somerville, R.P. Randomized selection design trial evaluating CD8+-enriched versus unselected tumor-infiltrating lymphocytes for adoptive cell therapy for patients with melanoma. *J. Clin. Oncol.* 31: 2152–2159, 2013.

Dunn, G.P., Old, L.J., Schreiber, R.D. The immunobiology of cancer immunosurveillance and immunoediting. *Immunity,* 21: 137–148, 2004.

Egen, J.G., Allison, J.P. Cytotoxic T lymphocyte antigen-4 accumulation in the immunological synapse is regulated by TCR signal strength. *Immunity,* 16: 23–35, 2002.

Eiraku, M., Watanabe, K., Matsuo-Takasaki, M., Kawada, M., Yonemura, S., Matsumura, M., Wataya, T., Nishiyama, A., Muguruma, K., Sasai, Y. Self-organized formation of polarized cortical tissues from ESCs and its active manipulation by extrinsic signals. *Cell Stem Cell.* 3: 519–532, 2008.

El-Khoueiry, A.B., Melero, I., Crocenzi, T.S. Phase I/II safety and antitumor activity of nivolumab in patients with advanced hepatocellular carcinoma (HCC). *J. Clin. Oncol.* 33: (suppl; abstr LBA101), 2015.

Emens, L.A., Braiteh, F.S., Cassier, P. Inhibition of PD-L1 by MPDL3280A leads to clinical activity in patients with metastatic triple-negative breast cancer (TNBC). AACR Annual Meeting; 18–22; Philadelphia, PA. Abstract 6317, 2015.

Freeman, G.J., Long, A.J., Iwai, Y., Bourque, K., Chernova, T., Nishimura, H,, Fitz, L.J., Malenkovich, N., Okazaki, T., Byrne, M.C., Horton, H.F., Fouser, L., Carter, L., Ling, V., Bowman, M.R., Carreno, B.M., Collins, M., Wood, C.R., Honjo, T. Engagement of the PD-1 immunoinhibitory receptor by a novel B7 family member leads to negative regulation of lymphocyte activation. *J. Exp. Med.* 192: 1027–1034, 2000.

Galluzzi, L., Senovilla, L., Vacchelli, E. Trial watch: Dendritic cell-based interventions for cancer therapy. *Oncoimmunology.* 1: 1111–1134, 2012.

Gil, M., Komorowski, M.P., Seshadri M. CXCL12/CXCR4 blockade by oncolytic virotherapy inhibits ovarian cancer growth by decreasing immunosuppression and targeting cancer-initiating cells. *J. Immunol.* 193: 5327–5337, 2014.

Gnjatic, S., Cai, Z., Viguier, M. Accumulation of the p53 protein allows recognition by human CTL of a wild-type p53 epitope presented by breast carcinomas and melanomas. *J. Immunol.* 160: 328–333, 1998.

Hamanishi, J., Mandai, M., Ikeda, T. Durable tumor remission in patients with platinum-resistant ovarian cancer receiving nivolumab. ASCO 2015, *J. Clin. Oncol.* 33, 2015 (suppl; abstr 5570).

Hamaï, A., Codogno, P., Mehrpour, M. Cancer stem cells and autophagy: Facts and perspectives. *J. Cancer Stem Cell Res.* 2: 1005, 2014.

Han, L., Shi, S., Gong, T., Zang, Z., Sun, X. Cancer stem cells: Therapeutic implications and perspectives in cancer therapy. *Sci. Direct.* 3: 65–75, 2013.

Hebrok, M. Hedgehog signaling in pancreas development. *Mech. Dev.* 120: 45–57, 2003.

Honda, T., Egen, J.G., Lammermann, T., Kastenmuller, W., Torabi-Parizi, P., Germain R.N. Tuning of antigen sensitivity by T cell receptor-dependent negative feedback controls T cell effector function in inflamed tissues. *Immunity,* 40: 235–247, 2014.

Huang, R.Y, Eppolito, C, Lele, S, Shrikant, P, Matsuzaki, J, Odunsi, K. LAG3 and PD1 co-inhibitory molecules collaborate to limit CD8+ T cell signaling and dampen antitumor immunity in a murine ovarian cancer model. *Oncotarget.* 6: 27359–27377, 2015.

Huch, M., Bonfanti, P., Boj, S.F., Sato, T., Loomans, C.J., Van de Wetering, M., Sojoodi, M., Li, V.S., Schuijers, J., Gracanin, A., Ringnalda, K., Begthel, H., Hamer, K., Mulder, J., van Es, J.H., de Koning, E., Vries, E.G.H., Heimberg, H. Clevers, H. Unlimited in vitro expansion of adult bi-potent pancreas progenitors through the Lgr5/R-spondin axis. *EMBO J.* 32: 2708–2721, 2013.

Hwang, W.T., Adams, S.F., Tahirovic, E. Prognostic significance of tumor-infiltrating T cells in ovarian cancer: A meta-analysis. *Gynecol. Oncol.* 124: 192–198, 2012.

Intlekofer, A.M., Thompson, C.B. At the bench: Preclinical rationale for CTLA-4 and PD-1 blockade as cancer immunotherapy. *J. Leukoc. Biol.* 94: 25–39, 2013.

Kantoff, P.W., Higano, C.S., Shore, N.D. Sipuleucel-T immunotherapy for castration-resistant prostate cancer. *N. Engl. J. Med.* 363: 411–22, 2010.

Karthaus, W.R., Iaquinta, P.J., Drost, J., Gracanin, A., Van Boxtel, R., Wongvipat, J., Dowling, C.M., Gao, D., Begthel, H., Sachs, N., Vries, R.G.J., Cuppen, E.,

Chen, Y., Sawyers, C.L., Clevers, H.C. Identification of multipotent luminal progenitor cells in human prostate organoid cultures. *Cell.* 159: 163–175, 2014.

Keir, M.E.,Butte, M.J., Freeman, G.J., Sharpe. A.H. PD-1 and its ligands in tolerance and immunity. *Annu. Rev. Immunol.* 26: 677–704, 2008.

Kim, M.S., Pinto, S.M., Getnet, D., Nirujogi. R,S., Manda, S.S., Chaerkady, R., Madugundu, A.K., Kelkar, D.S., Isserlin, R., Jain, S., Thomas, J.K., Muthusamy, B., Leal-Rojas, P., Kumar, P., Sahasrabuddhe, N.A., Balakrishnan, L., Advani, J., George, B., Renuse, S., Selvan, L.D., Patil, A.H., Nanjappa, V., Radhakrishnan, A., Prasad, S., Subbannayya, T., Raju, R., Kumar, M., Sreenivasamurthy, S.K., Marimuthu, A., Sathe G.J., Chavan, S., Datta, K.K., Subbannayya, Y., Sahu, A., Yelamanchi, S.D., Jayaram, S., Rajagopalan, P., Sharma, J., Murthy, K.R., Syed, N., Goel, R., Khan, A.A., Ahmad, S., Dey, G., Mudga, I K., Chatterjee, A., Huang, T.C., Zhong, J., Wu, X., Shaw, P.G., Freed, D., Zahari, M.S., Mukherjee, K.K., Shankar, S., Mahadevan, A., Lam, H., Mitchell, C.J., Shankar, S.K., Satishchandra, P., Schroeder, J.T., Sirdeshmukh,, R, Maitra, A., Leach, S.D., Drake, C.G., Halushka, M.K., Prasad, T.S., Hruban, R.H., Kerr, C.L., Bader, G.D., Iacobuzio-Donahue, C., Gowda, H., Pandey, A. S draft map of the human proteome. *Nature,* **509**: 575–581, 2014.

Klaus, A., Birchmeier, W. Wnt signalling and its impact on development and cancer. *Nat. Rev. Cancer.* **8**: 387–398, 2008.

Kumar, D., Shankar, S., Srivastava, R.K. Understanding the biological functions and therapeutic potentials of stem cells and cancer stem cells: Where are we? *J. Cancer Stem Cell Res.* **2**: 1002, 2014.

Lancaster, M.A., Knoblich, J.A. Organogenesis in a dish: Modeling development and disease using organoid technologies. *Science,* **345**: 1247125, 2014.

Lancaster, M.A., Renner, M., Martin, C.A., Wenzel, D., Bicknell, L.S., Hurles, M.E., Homfray, T., Penninger, J.M., Jackson, A.P., Knoblich. J.A. Cerebral organoids model human brain development and microcephaly. *Nature,* **501**: 373–379, 2013.

Larkin, J., Chiarion-Sileni, V., Gonzalez, R. Combined nivolumab and ipilimumab or monotherapy in untreated melanoma. *N. Engl. J. Med.* **373**: 23–34, 2015.

Laurent, L.C., Ulitsky, I., Slavin, I., Tran, H., Schork, A., Morey, R., Lynch, C., Harness, J.V., Lee, S., Barrero, M.J., Ku, S., Martynova, M., Semechkin, R., Galat, V., Gottesfeld, J., Izpisua Belmonte, J.C., Murry, C., Keirstead, H.S., Park, H.S., Schmidt, U., Laslett, A.L., Muller, F.J., Nievergelt, C.M., Shamir, R., Loring, J.F. Dynamic changes in the copy number of pluripotency and cell proliferation genes in human ESCs and iPSCs during reprogramming and time in culture. *Cell Stem Cell.* **8**: 106–118, 2011.

Ledford, H. US cancer institute overhauls cell lines. *Nature,* 530: 391, 2016.

Li, L., Borodyansky, L., Yang, Y. Genomic instability en route to and from cancer stem cells. *Cell Cycle*, **8**: 1000–1002, 2009.

Linsley, P.S., Ledbetter, J.A. The role of the CD28 receptor during T cell responses to antigen. *Annu. Rev. Immunol.* **11**: 191–212, 1993.

MacKie, R.M., Reid, R., Junor, B. Fatal melanoma transferred in a donated kidney 16 years after melanoma surgery. *N. Engl. J. Med.* **348**: 567–568, 2003.

Marincola, F.M., Jaffee, E.M., Hicklin, D.J., Ferrone, S. Escape of human solid tumors from T-cell recognition: Molecular mechanisms and functional significance. *Adv. Immunol.* **74**: 181–273, 2000.

Macleod, M. Why animal research needs to improve. *Nature,* **477**: 511, 2011.

Medema, J.P. Cancer stem cells: The challenges ahead. *Nat. Cell. Biol.* **15**: 338–344, 2013.

Melief, C.J., Van der Burg, S.H. Immunotherapy of established (pre)malignant disease by synthetic long peptide vaccines. *Nat. Rev. Cancer.* **8**: 351–360, 2008.

Motzer, R.J., Escudier, B., McDermott, D.D., George, S., Hammers, H.J., Srinivas, S., Tykodi, S.S., Sosman, J.A., Procopio, G., Plimack, E.R., Castellano, D., Choueiri, T.K., Gurney, H., Donskov, F., Bono, P., Wagstaff, J., Gauler, T.C., Ueda, T., Tomita, Y., Schutz, F.A., Kollmannsberger, C., Larkin, J., Ravaud, A., Simon, J.S., Xu, L.A., Waxman, I.M,, Sharma, P. Nivolumab versus everolimus in advanced renal-cell carcinoma. *N. Engl. J. Med.* **373**: 1803–1813, 2015.

Munn, D.H., Sharma, M.D., Mellor, A.L. Ligation of B7-1/B7-2 by human CD4+ T cells triggers indoleamine 2,3-dioxygenase activity in dendritic cells. *J. Immunol.* **172**: 4100–4110, 2014.

Naluai, A.T., Nilsson, S., Samuelsson, L., Gudjonsdottir, A.H., Ascher, H., Ek, J., Hallberg, B., Kristiansson, B., Martinsson, T., Nerman, O., Sollid, L.M., Wahlström, J. The CTLA4/CD28 gene region on chromosome 2q33 confers susceptibility to celiac disease in a way possibly distinct from that of type 1 diabetes and other chronic inflammatory disorders. *Tissue Antigens.* **56**: 350–355, 2000.

Nelson, B. Cultural divide. *Nature.* **509**: 152–154, 2014.

Nishimura, H., Okazaki, T., Tanaka, Y., Nakatani, K., Hara, M., Matsumori, A., Sasayama, S., Mizoguchi, A., Hiai, H., Minato, N., Honjo, T. Autoimmune dilated cardiomyopathy in PD-1 receptor-deficient mice. *Science,* **291**: 319–322, 2001.

Nozawa, H., Tadakuma, T., Ono, T., Sato, M., Hiroi, S., Masumoto, K., Sato, Y. Small interfering RNA targeting epidermal growth factor receptor enhances chemosensitivity to cisplatin, 5-fluorouracil and docetaxel in head and neck squamous cell carcinoma. *Cancer Sci.* **97**: 1115–1124, 2006.

Old, L.J. Cancer/testis (CT) antigens — a new link between gametogenesis and cancer. *Cancer Immun.* **1**: 1, 2001.

Old, L.J. Cancer vaccines: An overview. *Cancer Immun.* **8**(Suppl. 1): 1, 2008.

Pastrana, E., Silva-Vargas, V., Doetsch, F. Eyes wide open: A critical review of sphere-formation as an assay for stem cells. *Cell Stem Cell.* **8**: 486–498, 2011.

Purton, L.E., Scadden, D.T. Limiting factors in murine hematopoietic stem cell assays. *Cell Stem Cell.* **1**: 263, 2007.

Plimack, E.R., Bellmunt, J., Gupta, S. Pembrolizumab (MK-3475) for advanced urothelial cancer: Updated results and biomarker analysis from KEYNOTE-012. ASCO 2015, *J. Clin. Oncol.* **33**, 2015 (suppl; abstr 4502).

Przyborski, S.A. Differentiation of human embryonic stem cells after transplantation in immune-deficient mice. *Stem Cells.* **23**: 1242–1250, 2005.

Qiao, J., Kottke, T., Willmon, C. Purging metastases in lymphoid organs using a combination of antigen-nonspecific adoptive T cell therapy, oncolytic virotherapy and immunotherapy. *Nat. Med.* **14**: 37–44, 2008.

Qiao, J., Wang, H., Kottke, T. Loading of oncolytic vesicular stomatitis virus onto antigen-specific T cells enhances the efficacy of adoptive T-cell therapy of tumors. *Gene Ther.* **15**: 604–616, 2008.

Qiu, X., Wang, Z., Li, Y. Miao, Y., Ren, Y., Lua, Y. Characterization of sphere-forming cells with stem-like properties from the small cell lung cancer cell line H446. *Cancer Lett.* **323**: 161–170, 2012.

Qureshi, O.S., Zheng, Y., Nakamura, K., Attridge, K., Manzotti, C., Schmidt, E.M., Baker, J., Jeffery, L.E., Kaur, S., Briggs, Z., Hou, T.Z., Futter, C.E., Anderson, G., Walker, L.S., Sansom, D.M. Trans-endocytosis of CD80 and CD86: A molecular basis for the cell-extrinsic function of CTLA-4. *Science,* **332**: 600–603, 2011.

Robbins, P.F., Morgan, R.A., Feldman, S.A. Tumor regression in patients with metastatic synovial cell sarcoma and melanoma using genetically engineered lymphocytes reactive with NY-ESO-1. *J. Clin. Oncol.* **29**: 917–924, 2011.

Rosenberg, S.A., Restifo, N.P., Yang, J.C. Adoptive cell transfer: A clinical path to effective cancer immunotherapy. *Nat. Rev. Cancer.* **8**: 299–308, 2008.

Rudd, C.E., Taylor, A., Schneider, H. CD28 and CTLA-4 coreceptor expression and signal transduction. *Immunol. Rev.* **229**: 12–26, 2009.

Santucci-Pereira, J., Barton, M., Russo, J. Use of next generation sequencing in the identification of long non coding RNA as potential players in breast cancer prevention. *Trasncriptomic.* **2**: 1–5, 2014.

Sato, T., Vries, R.G., Snippert, H.J., Van de Wetering, M., Barker, N., Stange, D.E., Van Es, J.H., Abo, A., Kujala, P., Peters, P.J., Clevers, H. Single Lgr5 stem cells build crypt-villus structures in vitro without a mesenchymal niche. *Nature,* **459**: 262–265, 2009.

Savina, A., Jancic, C., Hugues, S. NOX2 controls phagosomal pH to regulate antigen processing during crosspresentation by dendritic cells. *Cell.* **126**: 205–218, 2006.

Sharma, H., Sen, S., Mraiggiò, L.M.L., Singh, N. Antisense-mediated downregulation of antiapoptotic proteins induces apoptosis and sensitises head and neck squamous cell carcinoma cells to chemotherapy. *Cancer Biol. Ther.* **4**: 720–727, 2005.

Schneider, H., Downey, J., Smith, A., Zinselmeyer, B.H., Rush, C., Brewer, J.M., Wei B., Hogg, N., Garside, P., Rudd, C.E. Reversal of the TCR stop signal by CTLA-4. *Science*, **313**: 1972–1975, 2006.

Scanlan, M.J., Chen, Y.T., Williamson, B. Characterization of human colon cancer antigens recognized by autologous antibodies. *Int. J. Cancer.* **76**: 652–658, 1998.

Shankaran, V., Ikeda, H., Bruce, A.T. IFNgamma and lymphocytes prevent primary tumour development and shape tumour immunogenicity. *Nature*, **410**: 1107–1111, 2001.

Simpson, A.J., Caballero, O.L., Jungbluth, A. Cancer/testis antigens, gametogenesis and cancer. *Nat. Rev. Cancer.* **5**: 615–625, 2005.

Singh, B.N., Fu, J., Srivastava, R.K., Shankar, S. Hedgehog signaling antagonist GDC-0449 (Vismodegib) inhibits pancreatic cancer stem cell characteristics: Molecular mechanisms. *PLoS ONE.* **6**: 27–30, 2011.

Snyder, A., Makarov, V., Merghoub, T. Genetic basis for clinical response to CTLA-4 blockade in melanoma. *N. Engl. J. Med.* **371**: 2189–2199, 2014.

Spence, J.R., Mayhew, C.N., Rankin, S.A., Kuhar, M.F., Vallance, J.E., Tolle, K., Hoskins, E.E., Kalinichenko, V.V., Wells, S.I., Zorn, A.M., Shroyer, N.F., Wells, J.M. Directed differentiation of human pluripotent stem cells into intestinal tissue *in vitro*. *Nature*, **470**: 105–109, 2011.

Takasato, M., Er, P.X., Becroft, M., Vanslambrouck, J.M., Stanley, E.G., Elefanty, A.G., Little, M.H. Directing human embryonic stem cell differentiation towards a renal lineage generates a self-organizing kidney. *Nat. Cell Biol.* **16**: 118–126, 2014.

Takebe, T., Sekine, K., Enomura, M., Koike, H., Kimura, M., Ogaeri, T., Zhang, R.R., Ueno, Y., Zheng, Y.W., Koike, N., Aoyama, S., Adachi, Y., Taniguchi, H. Vascularized and functional human liver from an iPSC-derived organ bud transplant. *Nature*, **499**: 481–484, 2013.

Teft, W.A., Kirchhof, M.G., Madrenas, J. A molecular perspective of CTLA-4 function. *Annu. Rev. Immunol.* **24**: 65–97, 2006.

Tindle, R.W. Human papillomavirus vaccines for cervical cancer. *Curr. Opin. Immunol.* **8**: 643–650, 1996.

Tomasetti, C., Vogelstein, B. Cancer etiology: Variation in cancer risk among tissues can be explained by the number of stem cell divisions. *Science*, **347**: 78–81, 2015.

Tsuji, T., Matsuzaki, J., Kelly, M.P. Antibody-targeted NY-ESO-1 to mannose receptor or DEC-205 in vitro elicits dual human CD8+ and CD4+ T cell responses with broad antigen specificity. *J. Immunol.* **186**: 1218–1227, 2011.

Tumeh, P.C., Harview, C.L., Yearley, Y.H. PD-1 blockade induces responses by inhibiting adaptive immune resistance. *Nature*, **515**: 568–571, 2014.

Walter, S., Weinschenk, T., Stenzl, A. Multipeptide immune response to cancer vaccine IMA901 after single-dose cyclophosphamide associates with longer patient survival. *Nat. Med.* **18**: 1254–1261, 2012.

Wolfel, T., Hauer, M., Schneider, J. A p16INK4a-insensitive CDK4 mutant targeted by cytolytic T lymphocytes in a human melanoma. *Science*, **269**: 1281–1284, 1995.

Wu, X., Chen, H., Wang, X. Can lung cancer stem cells be targeted for therapies? *Cancer Treat Rev.* **38**: 580–588, 2012.

Wu, S., Powers, S., Zhu, W., Hannun, Y. A substantial contribution of extrinsic risk factors to cancer development. *Nature*, **529**: 43–47, 2016.

Further Reading

Dudley, M.E., Yang, J.C., Sherry, R. Adoptive cell therapy for patients with metastatic melanoma: Evaluation of intensive myeloablative chemoradiation preparative regimens. *J. Clin. Oncol.* **26**: 5233–5239, 2008.

Pannuti, A., Foreman, K., Rizzo, P., Osipo, C., Golde, T. Targeting notch to target cancer stem cells. *Clini. Cancer Res.* **16**: 3141–3152, 2010.

Scanlan, M.J., Altorki, N.K., Gure, A.O. Expression of cancer-testis antigens in lung cancer: Definition of bromodomain testis-specific gene (BRDT) as a new CT gene, CT9. *Cancer Lett.* **150**: 155–164, 2000.

Wang, Z.W., Li, Y.W., Banerjee, S., Sarkar, F.H. Exploitation of the notch signaling pathway as a novel target for cancer therapy. *Anticancer Res.* **28**: 3621–3630, 2008.

Chapter 3

The Communality of Science in Cancer Research

3.1. Introduction

The reality is that lay people today know more about cancer than at any other time in human history. This is mainly due to the open source of knowledge provided by the Internet, advances in communications technology, and the rapidly expanding global fiber optic network. There is also significant effort to bring together scientists, educators, and lay people in national and international scientific meetings, creating a unique situation in which knowledge of the interaction between the environment and health — particularly its impact on cancer incidence and the new approaches of diagnosis, treatment, and prevention — is more treasured by the people around the world than ever before. This global movement of diverse societies to value knowledge in these areas has also been observed in other issues pertinent to our survival as humans, such as finding energy sources that do not contribute to climate warming and developing agricultural methods that can feed the 9 billion people inhabiting our planet with less water and a smaller ecological footprint. This communality in knowledge and interest in innovations in cancer research have been also extended to other areas of endeavor, including providing access to public health measures and clean water and preserving our biodiversity, restoring degraded ecosystems, controlling population growth, and stimulating economic development. All these easily accessible sources of knowledge are creating an unprecedented sense of communality in cancer research by creating a tighter network of partnerships and interchange.

3.2. How the Communality of Science has Evolved in Japan

Promising young researchers need to choose the best fit for their continued training; they must select a place that will foster their independent career as cancer researchers. In some places, such as United States, there is a vast and varied array of options for up-and-coming researchers to choose from. However, in countries such as Japan young researchers do not have this freedom to choose. Japanese academia has a very strict hierarchy in which young researchers are sent to a place for training instead of choosing where they would like to be trained. However, this rigid Japanese academic system has been relaxed in the last 15 years by the creation of centers that are more open and attractive to young researchers, not only from Japan but from all over the world (Cyranoski, 2011). This new aperture will be a breath of fresh air for the new generation of cancer researchers.

3.3. How the Communality of Science has Evolved in Africa

In Africa, the number of people is growing faster than anywhere else: the population is predicted to swell from the current 1 billion to 3.5 billion by the turn of the century (Irikefe *et al.*, 2011). Therefore, scientific and technical advances — particularly those that draw on research performed on the African continent — will be central areas of research inquiry. Besides South Africa, the science output and education across sub-Saharan Africa is creating a great opportunity to integrate into the communality of science due to the tremendous needs in research education. This new generation of African scientists needs to face a culture where science and mathematics are not encouraged in general, and if they are, they are considered subjects only for boys; girls are dissuaded from pursuing science. In a promising initiative each member of the African Union pledged to spend 1% of its gross domestic product (GDP) on research and development (R&D) (Irikefe *et al.*, 2011). However, only Malawi, Uganda, and South Africa topped the 1% spending threshold in 2007; most member nations remained far from that mark even when the support from foreign donors was included. The basic problem is a lack of infrastructure in this region, including governments crippled by issues ranging from corruption to ineffective bureaucracy. Research facilities are poorly equipped, and science students get little practical research training because research centers are often separate from universities (Irikefe *et al.*, 2011).

It is difficult to imagine how these countries can be part of the communality of cancer research when they are facing more pressing issues such as poverty, rampant infectious diseases, and the lack of clean water and energy (Irikefe *et al.*, 2011). Although money is often cited as the main problem, it is also important to reverse the brain drain that is robbing Africa of its leading scientists and engineers. However, any solution will no doubt require government help. *Only people can help solve people problems, and scientists and engineers can help develop libraries, laboratories, and provide the know-how to do R&D.* Some of these ideas are already in practice in Uganda and Nigeria (Irikefe *et al.*, 2011); however, Uganda has one of the lowest densities of researchers among the most scientifically advanced nations in sub-Saharan Africa: just 25 researchers per million inhabitants.

Among sub-Saharan nations Kenyan science ranks third, behind South Africa and Nigeria. Most of the scientific work in Kenya takes place in government-owned research institutes that use extensive international collaborations. Among the most renowned is the Kenya Medical Research Institute (KEMRI), which has centers around the country and does basic research as well as develops drugs, vaccines, and products such as diagnostic kits for HIV. KEMRI has a budget of $37.5 million, with 45% coming from its international collaborators, including the Wellcome Trust, a London-based medical research charity (Irikefe *et al.*, 2011).

Tanzania ranks as one of Africa's top five countries in terms of immunology publications and also scores highly in social and environmental sciences. Senegal has 661 researchers per million inhabitants; the country is second only to South Africa in researcher density (Irikefe *et al.*, 2011). Senegal has also opened the African Institute for Mathematical Sciences in Mbour, on the coast south of the capital, Dakar. Like researchers in Tanzania, Senegal's receive more than 38% of their funding from abroad; with national funding scarce, it is difficult to sustain long-term research projects. Rwanda has improved many of its institutions and has had some success in reducing the prevalence of malaria and HIV/AIDS. There are plans to increase R&D spending, with a focus on both constructing and equipping science laboratories and health and agriculture research. The African Institute for Mathematical Sciences can be considered an important agglutinating factor in developing the scientific environment of Africa as it clusters African scientists in Europe, United States, and China with the new generation of scientists living on the African continent.

The support coming from other countries has undoubtedly helped Africa in both the short and long range. China has recently allowed more students to spend time in China, boosting collaborations. The number of African students in China rose by 40% to nearly 4000 between 2005 and 2006, and the trend has continued. China is also emerging as a science collaborator with Africa, a role traditionally occupied by the United States and Europe. However, China's science investment is still outpaced by US and European funding.

Science, and specifically cancer research, will require some time to reach maturity in Africa as well as other poor countries in the world, but only time will tell the true story of cancer research's development in Africa. As recently as 1860, Japan was in medieval times compared with Europe and the United States, and the tremendous jump of that country to the 21st century was unprecedented. But Japan's progress clearly illustrates it is important to believe in the communality of science and the emerging global need of intellectual enrichment that will eventually permeate all the countries in the world (Irikefe *et al.*, 2011).

3.4. How the Communality of Science has Evolved in India

India has nearly 1.3 billion people and in a generation's time, it will be the most populous nation on the earth (Padma, 2015). Besides this increase in population, its economy is steadily growing and its scientific production, measured by the number of publications, has quadrupled in the last 13 years. However, despite these achievements, India is not yet a major player in world science; this is measured by the number of citations of their publications, plus the relatively low number of scientists in relation to the population. Also relevant is the low investment in R&D and the number of patents per capita, which is less than other nations. However on the positive side, new, important centers are emerging, and the number of women in the science force is increasing. Another promising development is the number of young investigators trained in the United States who are returning to India with the conviction that high-quality research can be carried out in the country. This is supported by the government and India's thriving pharmaceutical industry, which produces low-cost medications that are desperately needed by the developing world (Padma, 2015). But India is battling criticism over the quality of some of its pharmaceuticals and "while some Indian companies meet US product quality standards, others have been found to lack sufficient controls and systems to assure drug quality both of finished product and active ingredients" (Padma, 2015).

There is no doubt that India is facing challenging health and infrastructure issues — for example, high rates of tuberculosis and maternal deaths, plus the lack of electricity for one-quarter of its citizens — that will require a robust science and technology sector to supply the needed energy, food, health care, jobs, and growth. A 2014 report from the World Economic Forum and Harvard School of Public Health estimates that noncommunicable diseases and mental illness could cost India $4.58 trillion by 2030. All of this means that in order for India to succeed in the scientific research endeavor, and mainly in cancer research, they need to first solve their systemic problems as a nation.

In my own perspective, India has an advantage in its inherited English language that allows its citizens to easily communicate with the advanced societies of the West. This language advantage is significant when compared with the struggle that Spanish-, Arabic-, Japanese-, and Chinese-speaking people need to overcome to communicate in English; language is the main barrier that separates people in general, scientists no less. The commonality of science is based in the language and English is the language of science, therefore, India has a tremendous advantage that in my vision they are using quite well. While it is true that they are facing significant problems from overpopulation, they have the will to emerge as a big player in the community of research and also in cancer research.

3.5. How the Communality of Science has Evolved in South America

South America comprises many countries that share two main languages, Spanish and Portuguese, and a common origin, conquest and colonization by the inhabitants of the Iberic Peninsula in the 15th century. But other than these similarities, their histories are divergent and different. For example, Brazil was born from Portugal, which favored slavery; in contrast those nations conquered and colonized by Spain were not as heavily influenced by slavery; in fact marriage between Spaniards and indigenous people was common in countries such as Ecuador and Peru. The same process of conquering and intermixing took place in Mexico and many of the Central American countries. What makes each country different today is the demographic distribution of their people, influenced in great part by the immigration policies of each nation. For example, almost 60% of Argentina's population is of Italian descent, whereas in Brazil, Venezuela, and Colombia the number is less than 5%. Therefore, the communality of science in South America cannot be properly analyzed without taking at least these factors into consideration.

The main problem is that if the common metric of productivity, i.e., the number of publications, is used in South America, it would still fall short of what would be expected. Research quality has not kept pace with rising output, and the continent's research papers still struggle to attract citations from the rest of the world. According to Richard Van Noorden (2014), there are huge inequalities across the region, too: e.g., Brazil dominates the publication record, whereas Chile takes pole position in the patent landscape, and Argentina scores highly in terms of the proportion of its population working in science. In Argentina, the number of science doctorates jumped nearly 10-fold between 2000 and 2010, and Peruvian scientists tripled the tally of articles they produced over the same period. In addition, science funding is increasing in most countries. But these indicators can also be misleading about South America's place in the world. For example, the article published by Richard Van Noorden (2014) in *Nature* indicated that Brazil produced 46,303 publications in comparison with Argentina's 9,337; but this metric does not consider that Brazil has 300 million inhabitants whereas Argentina has only 33 million. If we recalculate the metric, Brazil produces only 154 publications per million inhabitants and Argentina produces 282 publications per million inhabitants — that is almost double in productivity, and also the impact of Argentinian science is significantly higher than that of all the other counties of South America. Meaning that the way numbers are presented (Van Noorden,

2014) can be deceiving. Scientific research in Colombia, Venezuela, and Peru has been in arrested development due to the sociopolitical upheaval that those countries have suffered for the last 20 years. It is logical that when these countries stabilize themselves science will emerge again.

There are also significant differences in the way that scientific research has emerged in Brazil and in Argentina. Brazilian science has emerged basically by the creation of FAPESP (Fundação de Amparo à Pesquisa do Estado de São Paulo), a state agency that promotes research and education. Although this agency has its stronghold in Sao Paulo, other states of Brazil are using this it as a model. The interesting aspect of FAPESP is that Sao Paulo is one of the few states in the world where support of research is linked directly to GDP (Miranda, 2014). By comparison, Argentina is the South American country with more tradition in scientific research as well as in cancer research. Scientific research started in the early 19th century and organized in the middle of the 20th century with Bernardo Houssay and Luis Leloir, 20th-century Nobel laureates. The scientific institutions they founded led to generations of disciples that continue to do the science of today. The reality of Argentina is that it produces more scientists per capita than the country can absorb and therefore migration of brains has been characteristic of Argentina since the middle of the 20th century. In that sense, the Argentinian Phenomenon is quite unique in Latin America. In a certain way the connection of Argentina with Europe since the former nation's creation in 1810 has made this country's population very comfortable with not only migrating to Europe but also training there. In addition, the political and economic evolution of Argentina during the last 30 years has been unstable, making it less appealing to the United States. Conversely, the faster growing economy of Brazil has put that country more in the vision of the USA scientists, as well as in the news. It seems that this connection has been lost by many articles in the literature.

3.6. Information Influencing the Communality of Science

We are living in a golden age of information and a dark age of ill-informed opinions; traditional and start-up media alike frantically cast their nets to engage audiences for more than a nanosecond. The relevant question is how useful will social media be in the future of cancer research, and more importantly, what impact will it have in the future of our species. I know these are tough questions and that I will be unable to provide the final answers, but on the other hand, these difficult questions may help to reshape what we are expecting from the information era.

One thing that is clear is that the information we receive is targeted to shape the marketplace and determine how we react to the offerings and demands of our society. This in itself is not bad because in a free economy the consumer or society needs to be informed. The information is coming from different paths such as newspapers, magazines, websites, and even scientific literature. All of these pipelines create the information that we receive, and we search for it because we believe that if we have that information it will make us smarter. This is particularly useful in medicine; there are multiple websites with valuable information on the definition of a disease, diagnosis, prognosis, and treatment, as well as numerous links to specialists in the field. The conclusion is that the overload of available information needs to be filtered by the target, meaning the person that is the end point. Most of the websites related to health are intended to serve, and they offer different levels of complexity that are easily adjusted to the educational background of the recipient. There are thousands of useful and educational web pages created by private and government institutions, and some of them are not selling anything, only offering educational material intended for the well-being of the user. Everybody feels smarter after reading up on a topic, but the next step is to learn how to use the information provided wisely and how to discern if the information acquired is correct and applicable.

The new social media has not been overlooked by scientists, and whereas it is true that some branches of science use them more often than others (Morello, 2015), there is a new current of popularity that forces its use. Social media has enabled an increasingly public discussion about the persistent problem of sexism in science (Morello, 2015) and Twitter conversations, for example, can help to amplify the voices of people who are not powerful by conventional measures. Thus it could be an important outlet for discussing concerns not only in the general public but also among scientists. Although it is used by the younger generation, the older generation of scientists in cancer research is far less inclined to use it.

The problem with social media is how to measure or predict the consequences. For example, it is difficult to determine how a handful of results will transform into an Internet firestorm, or what gives a hashtag staying power. It has been shown that stories that inspire intense positive emotions, such as awe or amusement, are the most likely to go viral; anger, anxiety, and other strong negative feelings also propel articles to wide readership, but sadness seems to reduce the chance that a reader will share a story with others (Morello, 2015). Researchers tracking the rise of social media are trying to understand whether intense discussions online translate into real-world change. The difficulty lies in deciding how to measure such effects (Morello, 2015).

This brings us to the final realization that great discoveries in science are currently announced by the "proper" channels, usually peer-reviewed publications in high-impact journals. Once these discoveries are in the system, then social media can discuss the findings, but before that crucial moment it is only a rumor of something coming, nothing that could be seriously considered. On the other hand, the new cadre of cancer research scientists can change this present paradigm and choose less stringent criteria, something many system biologists would like to see (see Chap. 2).

The Internet has undeniably unified the interaction of all researchers. This interconnectivity is one of the highest achievements of the "new millennium" and will be the main link among scientists. I will not expand myself in this subject because it has been already discussed in the scientific and lay press so much that it is difficult to even cite all of them.

Other popular social media are Myspace and Facebook, which many teens use to post prolific reports on their moods, drinking, drug use, and sex lives. This audience is important because it could be a target of medical attention and an arena in which educational tools could be implemented. The reason why this audience is important is because they are among the least likely people to see a doctor, and three-quarters of teens and young adults are on a social networking site, according to the Pew Internet & American Life Project, additionally many use social media to disclose the very behaviors that are likely to put them in danger. The problem is how to teach without intruding into the privacy of the people who are discussing their personal lives. Thus the most important question is how valuable would this intervention or effort from the medical profession be. This is an area that requires study, and the new cadre of cancer researchers needs to figure out how to use powerful social media to educate young adults in the prevention of cancer. But first researchers must acquire the know-how to prevent cancer and establish clear guidelines that people can follow. Unfortunately at this time, there are no clear and effective guidelines available.

Finding answers to questions about health in this vast information network is becoming increasingly difficult, even for the experts. One of the main reasons is that in this communality of data-sharing, the data exist but frequently are not meshed together because they have been collected differently or stored in databases and infrastructures that are not communicating with each other. There are divergent opinions on the use of meta-analysis to obtain valid conclusions, whereas others consider that the combination and analysis of large data sets are vulnerable to spurious correlations, such as genomic or electronic medical record data.

A new issue that has emerged is the concept of transparency of scientific research; many research findings are not reproducible, creating a sense of malaise that needs to be overcome. This issue of transparency was discussed in a recent article in *Nature* by Stephen Lewandowsky and Dorothy Bishop (2016) explaining how the research community should protect its members from harassment, while encouraging the openness that has become essential to science. Therefore, it is clear that in this time of anxiety for information the new cadre of cancer researchers need to address this problem.

3.7. Globalization and Science

Cancer research is in the process of globalization; this is easily seen in national as well as international cancer meetings. It is evident that there is a worldwide exchange of people, goods, money, information, and ideas that without doubt are overall beneficial. However, in a fascinating article, Helbing discusses the idea that besides the benefits of globalization the underlying networks have created pathways along which dangerous and damaging events can spread rapidly and globally; this has increased systemic risks. It is important to consider this point of view because usually only the benefits of globalization are emphasized; however, in this article Helbing provides evidence of the dangerous situation in which we are immersed. He summarizes that due to globalization the world is already facing some of the consequences: global problems such as fiscal and economic crises, global migration, and an explosive mix of incompatible interests and cultures, coming along with social unrests, international and civil wars, and global terrorism. This danger is explained by the networks that have been created and their interdependency. Helbing elaborates that our society is entering in a new era — the era of a global information society, characterized by increasing interdependency, interconnectivity, and complexity, and a life in which the real and digital world can no longer be separated. Helbing speculates that as interactions between components become "strong", the behavior of system components may seriously alter or impair the functionality or operation of other components. However, probabilistic cascade effects in real-life systems are often hard to identify, understand, and map. One example is what happens in a crowd-control situation and, as Helbing (2013) explains, the interaction strength increases with the crowd density, as people come closer together. When the density becomes too high, inadvertent contact forces are transferred from one body to another and add up. The resulting forces vary significantly in direction and size, pushing people around, and creating a phenomenon called "crowd quake". Turbulent waves cause people to stumble, and others fall over them in an often fatal domino effect. If people do not manage to get back on their feet quickly enough, they are likely to suffocate. In many cases, the instability is created not by foolish or malicious individual actions, but by the unavoidable amplification of small fluctuations above a critical density threshold. Consequently, crowd disasters cannot simply be evaded by policing, aimed at imposing "better behavior". Another example that Helbing elaborates is the following:

> the global financial meltdown due the massive trade in financial derivatives that created mega-catastrophic risks for the economy. The overall volume of credit default swaps and other financial derivatives had grown to several times the world gross domestic product. The collapse was produced even though everyone seemed to be doing their own job properly on its own merit. And according to standard measures of success, they were often doing it well. The failure was foreseeing how this

would collectively add up to a series of interconnected imbalances. Individual risks may rightly have been viewed as small, but the risk to the system as a whole was vast. (Helbing, 2013)

These examples are a clear indication that the problem of globalization is not in its concept but in the way that we are managing the system that allows globalization to take place. All this interconnectivity has created a gigantic database with redundancy in the data collected; in theory this should protect the individual but at the same time they have small margin for error. In addition, to be effective the networks created by these databases should be interconnected, and new advances and information systems that are added to the original system of data need to be created. All this complexity makes the system vulnerable.

Helbing proposes a set of suggestions for this hyper-risk but he is not fully convinced that there is a solution at hand. Among his suggestions for reducing hyper-risks is limitation of the system size by establishing upper bounds to the possible scale of disaster. He explains that in order to avoid a cascade effect we should construct fuses such as those in electrical circuits that serve to avoid large-scale damage of local overloads. A further principle that he proposes is to incorporate mechanisms that produce a manageable state. For example, if the system dynamics unfold so rapidly that there is a danger of losing control, one could slow it down by introducing frictional effects (such as a financial transaction fee that kicks in when financial markets drop). Also note that dynamical processes in a system can desynchronize if the control variables change too quickly relative to the time scale on which the governed components can adjust. For example, stable hierarchical systems typically change slowly on the top and much quicker on the lower levels. If the influence of the top on the bottom levels becomes too strong, this may impair the functionality and self-organization of the hierarchical structure. In addition, reducing connectivity may serve to decrease the coupling strength in the system. This implies a change from a dense to a sparser network, which can reduce contagious spreading effects. In fact, sparse networks seem to be characteristic for ecological systems. As logical as the above safety principles may sound, these precautions have often been neglected in the design and operation of strongly coupled, complex systems such as the world financial system (Helbing, 2013).

We can state that it is probably more expensive to make a computer-ized tomography (CT) scan than sequence the entire genome of a single individual. But the big difference is that we know how to interpret a CT scan and provide a diagnosis that can lead to a cure; we do not yet know how to figure out what those billions of bases and thousands of genes tell us about a particular disease. These big data efforts offer huge challenges, from creating analytic tools and solving scientific puzzles to accessing mil-lions of gigabytes of data and overcoming barriers to accessing patients' health records. As discussed in Chap. 2, genomic variants are one of the main data obtained in the genomic analysis that might reveal a tumor's therapeutic weak points or resistance to a given drug, and that is one of the main topics to be discerned by the new cadre of cancer researchers. National as well as international programs have been launched to under-stand the billions of data generated by the genomic sequence profile of thousands of people. The amount of data is so great that some researchers worry that the flood of information could overwhelm the computational pipelines needed for analysis and generate unprecedented demand for storage. It has been estimated that the output from genomics may soon dwarf data collected in YouTube (Eisenstein, 2015). However, the most important concern is not the storage but the clinical utility of that infor-mation; this has been a main factor in the decision to focus on the exome instead of the full genome, reducing by almost a hundredfold the amount of data and computational space needed. The negative part of this approach is that the exome, which is a subset of sequences containing protein-coding genes, only provides a fraction of the genomic variants.

A well-written article published in *Nature* by Eisenstein (2015) discusses the problems and possible solutions for this data storage and analysis. Basically the generation of data even from exomes requires 40 petabytes, or 40 million gigabytes, each year, and even that storage is a problem. Yet the major issue is that the analysis of these genomic variants can make sense at the clinical level but become exponentially more complicated when different combinations are taken into consideration. As discussed earlier, this networking of complexities is the main point that could make the system fail; therefore, a significant amount of effort and innovation in research strategies needs to be invested by the new cadre of cancer researchers to overcome this problem. This kind of safety net has not yet been constructed in any of the complex issues like finance or meteorology, to give some examples. The additional compound problem of the genomic data, as has been partially discussed in Chap. 2, is the personal data of each patient that is associated with the genomic or exome information. The solution must not only include regulation of how to have access to the data but also build a safety brake into the system that protects the data from intruders and the cascade effect of the complexity of the networks under study. Another challenging issue for the new cadre of cancer researchers is how to feasibly make the information accessible to the oncologist so he or she might determine, based on the genomic or exome profile, what approach makes more sense to use and what the outcome in response or secondary effect will be. The oncologist can predict this in percent basis but not on an individual basis.

In summary, the lesson that we are learning is that we have created big data that make it difficult to separate reliable information from ambiguous or incorrect data. The best example is the accumulation of millions of data points regarding the genomic pathways and the genes involved in the process and yet we still do not have a clear picture of how all the dots are connected. The question that we as cancer researchers must ask ourselves is that if we have all the data and the most powerful computer, can we predict the best treatment and the best outcome for a cancer patient? And the reality is that we can only predict response over a short period of time and only in the probabilistic sense, with a limited success concerning complications or unexpected events. If we extrapolate this to the global system, it is clear that we have created too many interactions that are making our human race significantly more vulnerable. We must hope that we keep working on these issues and develop shortcuts that allow us to stop the cascade effects of a failure in the interconnected networks. The metastatic spread of cancer makes the disease global and stopping the process of interconnected networks is a major challenge that we as cancer researchers must face.

References

Cyranosky, D. Okinawa goes recruiting. *Nature*, **17**: 551, 2011.
Eisenstein, M. The power of petabytes. *Nature*, **527**: S2–S4, 2015.
Helbingl, D. Globally networked risks and how to respond. *Nature*, **197**: 51–59, 2013.

Irikefe, V., Vaidyan, G., Nordling, L., Twahirwa, A., Nakkazi, E., Monastersky, R. The view from the front line. *Nature*, **474**: 556–559, 2011.

Lewandowsky, S., Bishop, D. Don't let transparency damage science. *Nature*, **529**: 459–461, 2016.

Miranda, G. Sao Paulo's heavy hitter. *Nature*, **510**: 205, 2014.

Morello, L. Social media is shaking up how scientists talk about sexism and gender issues. *Nature*, **527**: 149–151, 2015.

Padma, T.V. India's science test. *Nature*, **521**: 144–147, 2015.

Van Noorden, R. South America by the numbers. *Nature*, **510**: 202–203, 2014.

Chapter 4

The Economic Basis of Cancer Research

4.1. Introduction

It is often advertised in the press that the budget for medical research will be increased by several billion dollars. If this happens it would be really good news. But not all of this promised funding is expected to be effective, and it is common here in the United States as well as in Europe to see that ambitious research programs are dimmed when the overall budget is reduced. Economic fluctuations affect the budget allocated to medical research, and those places or countries that are lagging behind the larger center are affected the most.

4.2. The Use of Metrics to Manage the Distribution of Research Funds

In an ideal world of scientific research, each applicant for research funds would be evaluated on the originality of the idea proposed and the impact that this research would have in the overall field of cancer research. However, that is not the case. Instead, grant reviewers sweat through hundreds of applications; often they only have enough time to give each submission a cursory glance as they try to apply metrics that are defined as quantifiable measures of scientists' achievements, such as total citation count and the h-index, a measure of both the quality and quantity of papers (a scientist has an h-index of 12 if he or she has published 12 papers that have each received at least 12 citations). Naturally, many scientists object to such cold quantification of their contributions. Plus, all metrics have obvious flaws — a paper may gather many citations not because of its importance, but because it is in a large field that publishes frequently, and so generates more opportunities for citations. Review articles, which may not add much to the research, count the same as original research papers, which contribute a great deal. All existing metrics capture only what a scientist has done, not what he or she might be capable of. This limitation is described in a 2012 paper published in *Nature*: "Even we recognize there should be a better way … our peer review system is based on these metrics and I wonder if the reason why we are continuously using this practice is not forced by the general pressure that exists in the community on the scarcity of economic resources and that the metrics must in a certain way help to distribute the funds using this criteria".

However, we find that although we are advancing in our knowledge in cancer research, we are not objectively evaluating original ideas and when original ideas do emerge, the scientists that have the knack to think out of the box are restricted in their ability to get funds, as determined by applying the metrics. We have created a vicious circle in which, spurred by the intention to measure productivity, we are not rewarding potentially novel ideas and original innovative approaches that could change the cancer research field.

4.3. Looking for Funding: The Main Driving Force in Cancer Research

In a simplistic way, we can say that money is the combustion engine of the research enterprise. The funds are mostly coming from the National Institutes of Health (NIH), and the shift in the US funding landscape is clearly explained in an article published in *Nature* by Jennifer Couzin-Frankel (2014). She writes that over the past 20 years, federal investment in R&D as a share of the gross domestic product has fluctuated above and below 1%, and now stands a bit under it. Biology has long been a favored child of funders, its allure growing with time. Today, roughly two thirds of federal R&D money at universities goes to the life sciences, about 10% more than in the early 1970s. Industry spending also increased in the 1980s and 1990s, and now provides about 7% of the R&D dollars that flow to universities. At the same time, NIH's budget has sustained wild swings that many economists say make for an inefficient research enterprise. Between 1998 and 2003, the agency's budget doubled, from less than $14 billion to more than $27 billion. For the next 5 years it stayed largely flat. Then came an infusion of $10.4 billion in 2009, part of the federal stimulus plan to fight the recession — followed by a sizable bump downward in 2013, a 5% across-the-board cut from the sequester. Universities responded predictably to the budget doubling: They expanded, adding new buildings and filling them with staff members and trainees, who needed money of their own to thrive. Biomedical researchers had become dependent on annual budget increases of at least 6%. But as constraints take hold in biomedicine the instability is now on vivid display. On the one hand, the future looks a tad brighter: NIH's 2014 budget increased 3.5%, to $30 billion. But that will likely not be enough to sustain the community as it hopes. NIH's grant approval rate dropped below 17% last year, compared with about 30% in the late 1990s, and the average size of standard research grants fell for the first time in recent memory. Because NIH now approves less than one in five grant applications, scientists say they are spending more of their time submitting proposals — leaving less for the research needed to win grants in the first place.

The problem is that this economic upheaval has created a vacuum in the long-range incentives to succeed, because it becomes clear that early success is not an indication of continuous or sustained success in funding. It is possible to speculate different nightmares that will be created by this situation, like employment of highly skilled scientists who are now seeing their research endeavors cut off from the main stream, and we can continuously speculate on eventually reducing the number of scientists with the consequence of stagnation in our research and development (R&D) innovation. It is not my intention to describe all the worst-case outcomes but only to concentrate on the more *serious problems: particularly, the freedom of the independent investigator to pursue a research problem toward a solution.* It is not doubted that the government will continue to supply funds for R&D, but the problem is that the government enterprise is directed by the needs of society and therefore in a complex society, such as in the United States, our current problems are oriented toward distribution of the resources to the greatest number of people to fight disparity. Therefore, government funds will be available for solving those problems that address the immediate needs of society. Furthermore, those scientists who are able to adapt to solve problems determined by our immediate societal needs will be the successful ones. This is not fiction because in a certain way the success of Louis Pasteur was his ability to solve problems that were affecting the common man, local industries, and local farmers. From there great scientific discoveries emerged. But what happens with those basic research problems that are the driving force of the research endeavor of Western civilization? Has the freedom to pursue these research questions vanished from the fundable research interest? Are we entering into a period when our government agencies will only fund targetable research ideas, as determined by the needs of our society?

While I am not judging the wisdom of any of these outcomes determined by economic forces, I am concerned about the incentives that the new cadre of cancer researchers will have. Now the incentive to publish papers and write grant applications is the need to earn enough money to keep our research institutions alive. But if the incentives are not clear for the individual, the exercise of writing papers and grant applications, and maintaining the research institutions will be innocuous. If the incentive is profit, this could be a new way to pave an innovative type of science, like what happened with the Industrial Revolution; however, the risk is we can lose our path in the process.

4.4. The Reward for the Effort

Part of being trained as a research scientist is realizing that the path to accomplishing the research tasks leading to discovery is difficult and arduous. To write papers and grant proposals is part of the academic curriculum and everybody understands that this needs to be done and that it requires a compartmentalized mind. The scientific mind must be able to separate the day into time for doing research benchwork and the hours needed behind the computer cracking data, ideas, and projects. What was not expected, or at least was not part of the equation, is that the time expended to produce groundbreaking experiments has been significantly supplanted by the amount of time invested in writing grant applications. It is not unusual to spend all one's time submitting and resubmitting grant applications in the hope that they will pass the selection process and score funding, with the compounding problem that you have only two chances to pass in the NIH system of grant applications.

It is difficult for the world outside of cancer research to understand where the problem lies when they hear that the budget for cancer research is a huge one, but that it is never enough and young scientists have a hard time launching their careers and even the most senior scientists are trimming their research projects. The funding of scientific research, and cancer research in particular, is a compound problem. One of the issues is that most of the funds that come from the NIH need to be approved and spent on a yearly basis; however, the biggest issue is the growth of the scientific research population, because with that expansion comes a correlating demand for funding. For example, in 2000, the budget was approximately $13.7 billion and in 5 years it doubled to $27.1 billion, but it is still not enough. Part of these funds were for the construction of new buildings that were filled by researchers; in 1998, these researchers produced around 20,000 grant applications; in 2006, the number of applicants rose to 33,000. This rapid growth in the number of applications has increased the rejection rate, and consequently increased the number of resubmissions. From a bright and optimistic view this can be seen as an indication of a healthy intellectual pool ready to produce great and innovative ideas, but on the other hand, it is like being dressed for a party only to find the party has been cancelled. The equation is clear: the demand for funds for research has created a pressure on the federal budget, which cannot support those demands, and in addition it seems that the federal budget has little room to expand or even stabilize a minimal increase if research needs. A compounding problem is that most universities and medical institutions rely on NIH money for the bulk of scientists' salaries and overhead costs and are not set up to support faculty members in long term. In this high demand for funding the competition is ferocious and completely in the hands of grant reviewers — they are the only filter for grant allocations. Therefore, there is a need for outstanding study sessions with scientists who have not only the knowledge of how to do the science but also the maturity and experience to cast the right vote and the perfect score for those outstanding applications. In a society that requires so much different federal assistance, the most important priority is to preserve a high quality of cancer research by achieving outstanding study sessions that can do the best job to select which is the science that will make us progress to next new level.

4.5. Why a Healthy Economy in Cancer Research is Good for Society

Science investments should be measured neither in the short range nor in terms of the amount of money spent, but rather in terms of their capacity to produce great societal benefits for human beings. It is true that the benefits of basic scientific research are unpredictable and long term. Therefore, it is necessary to be proactive in educating the public as well as the policymakers that a small increase in basic research funding can affect a nation's capacity to solve social and economic problems. Scientists need to be aware that it is their duty to not allow a surgical approach to set priorities in the scientific research agenda. It is not doubted that science that results in immediate application to our health and social problems is needed, but there is also a great need to keep those research programs that are focused on the long term, not only those with short- to medium-term benefits.

A good equalizer in the funding of research is when the source is not only the government but also private philanthropists or foundations. In a certain way, the United States has been blessed by those philanthropists who have created extraordinary ways to cover different research endeavors. It is not the place of this chapter to enumerate those foundations or philanthropists but rather to provide a perspective of their role in cancer research or biomedical research in general. *Their effect cannot be measured in dollars because, more importantly, they are taking financial and scientific risks, filling the gaps left by government and industry, and helping the recipient to act fast in projects that otherwise would take a slower pace.* There is concern from certain sectors that these foundations and philanthropists retain a controlling power over the research endeavor and also the health policies derived from these discoveries, creating a "strings attached" situation as the funder keeps control over the direction of the research they pay for and demands a level of accountability that can make researchers uncomfortable. However, the benefits are palpable and whatever tax advantages that these organizations receive, they are filling a gap that is difficult to measure or deny. This appears to be a global trend, as money from nonprofit foundations and philanthropists is growing strikingly not only in United States and Great Britain but also in Germany (Wadman, 2007). In my perspective, these private foundations and philanthropists must be acknowledged and encouraged because if the money is funneled to mature research scientists and institutions that have proven to be of high quality, the research and benefits outcome will be easy to be measure.

4.6. Who Should be Funded in Times of Economic Constraint?

In the present climate of economic restraints on most of the academic research institutions, scientific researchers are judged on the amount of money they bring to their institutions (see Chap. 5). However, ludicrously enough, the research behind 30% of the work from Nobel laureates in medicine, for example, was done without direct funding (Tatsioni *et al.*, 2010). Therefore, this paradigm raises the following questions: Should the funds go to the outstanding projects? Or to the person who is the most promising? Or should funding be equally distributed among all the scientists?

Should the funds go to the outstanding projects? In analyzing a project, a good quality is that it is well written and crafted with enough information and preliminary data, and has an impeccable research approach with a very incisive explanation of the significance and innovation that the research will bring to the field. Based on this, the grant application is reviewed by the study session independently, whether the funds are from federal or nonfederal sources. Although there are exceptions in which a detailed application is not needed, at the end, a short or a long application must have this high quality of craftsmanship to be successful.

Should the funds be allocated to the person who is the most promising? This refers to those researchers who have had continuous production and achievements, which should be rewarded on the basis of their past and present works. There is not a written rule that this must be done, but in reality it happens. Either because the institutions want to reward this person for using funds obtained by private donations or because that individual does not run out of funds because he or she is the one with the best score in the study sessions. Probably the reason is that they have acquired or developed a high craftsmanship in writing grant applications.

Should funding be equally distributed among all the scientists? If we do it this egalitarian way in biomedical research, it is easy to conclude that each scientist will receive only a small amount, making it impossible to plan a large, complicated project or even moderate side experiments. However, in certain disciplines in which the laboratory cost is minimal, e.g., mathematics, small funding that allows travel or a piece of software or books will be adequate. By the same token, if biomedical researchers are able to secure a salary from their institutions, a small grant of $50,000 for supplies can also be a significant help creating preliminary data that will eventually support larger studies.

Because none of these ways to distribute funds for research are mutually exclusive, all of them have value, as indicated by Ioannidis (2013): "There is no good evidence on whether it is better to give fewer scientists more money or to distribute smaller amounts between more researchers". Experience indicates that the most innovative science is not always produced by the best-funded researchers, and in some cases it comes from an isolated individual with little support.

4.7. Does Patenting Improve the Economy of Scientific Cancer Research?

In contrast to the world of industry where patenting provides a significant economic advantage, there is doubt that the same benefit can be translated to biomedical research. This idea is reflected in US government policy establishing that unmodified DNA should not be patented because merely isolating something does not turn it into a human-made product. However, the concept of patenting research products or ideas is a hotly debated issue among scientists. Almost 36 years ago the Bayh-Dole Act established that US universities — not funding agencies — have the right to file and own intellectual property for inventions resulting from publicly funded research. The influence of this act has been shown in human genome research, indicating that the number of papers published and commercially available diagnostic tests are significantly higher because of the availability of data obtained by the government's Human Genomic Project (https://www.genome.gov/10001772/all-about-the--human-genome-project-hgp/) as opposed to data obtained by the private company Celera. The same amount of outcome has been observed in mouse genetic engineering, and the conclusions are that clear, simple, licensing guidelines and low-cost access to the mice enhanced follow-on research and prompted a burst of activity in novel areas. Mice strains derived from these technologies were cited at a rate 30% higher than expected levels for several years after the policy change (Furman *et al.*, 2010).

Making cell lines, gene sequences, engineered mice, or other types of reagents accessible through a trusted, open-access resource center increases their effect on research. It is not doubted that these regulations significantly decrease the incentive for the researchers; therefore, the new cadre of cancer researchers will need to find rules of practice that maximize the productivity of research in the long term — even if those rules cause today's researchers some inconvenience or loss of competitive edge (Furman *et al.*, 2010). It is important to establish equilibrium between rewards for the innovation provided by the individual researcher and the common good of that information to both present and future researchers.

4.8. The Role of Scientists in the Economy

When something at the societal level goes wrong, scientists are not always the first ones to be consulted. One example is the lead-contaminated drinking water in Flint, Michigan. This is a typical example in that scientists should have been consulted as soon as the problem was detected — not at the end of it. Another example of a situation in which scientists need to be involved right away is the development of the so-called circular economy. The concept of a circular economy is based on the idea that goods that are at the end of their service life can be turned into resources for others, closing loops in industrial ecosystems and minimizing waste (Stahel, 2016). It is obvious that excellence in metallurgical and chemical sciences is a precondition for the success of a circular economy; however, it is also important to consider biomedical knowledge in advance of development of a circular economy. The best examples are the significant amounts of discarded material in hospitals and in research laboratories, such as tissue culture plates, pipettes, and other disposable material (without considering antibodies, chemical reagents, and lastly scientific equipment). We in the biomedical field are used to a linear economy that is based on the concept of turning natural resources into base materials and products for sale through a series of value-adding steps. At the point of sale, ownership and liability for risks and waste pass to the buyer (who is now owner and user) (Stahel, 2016). Cleaning a glass bottle and using it again is preferable to using a disposable one. Recycling uniforms, industrial wipes, furniture, computers, and building components could be a significant way to reduce waste of resources. Understanding how to recycle the food discarded in hospitals every day would be a significant improvement; however, this requires developing and maintaining an infrastructure that would be intensive and expensive. One of the main reasons for this complexity is the fact that assurance of food availability requires some food surpluses (Papargyropoulou *et al.*, 2015) and the consequent food wasting creates a problem in disposal and decomposition in landfills with the subsequent environmental impact (Foley *et al.*, 2011). As well stated by Aschemann-Witzel (2016) the problem of food waste is a symptom of the currently unsustainable food supply. It is clear that food is part of the circular economy, and in the well-informed article by Stahel, all the different implications of the circular economy are given but there is still an unknown — the potential role of cancer researchers in this new vision of good management.

References

Aschemann-Witzel, J. Waste not, want not, emit less. *Science*, **352**: 408–409, 2016.

Couzin-Frankel, J. Chasing the money. *Science*, **344**: 24–25, 2014.

Foley, J.A., Ramankutty, N., Brauman, K.A., Cassidy, E.S., Gerber, J.S., Johnston, J.S., Mueller, N.D., O'Connell, C., Ray, D.K., West, P.C., Balzer, C., Bennett, E.M., Carpenter, S.R., Hill, C.J., Monfreda, C., Polasky, J.S., Rockström, J.J., Sheehan, J., Siebert, S., Tilman, D., and Zaks, D.P. Solutions for a cultivated planet. *Nature*, **478**: 337–342, 2011.

Furman, J.L., Murray, F., and Stern, S. More for the research dollar. *Nature*, **68**: 757–58, 2010.

Ioannidis, J.P.A. Fund people not projects. *Nature*, **477**: 529–531, 2013.

Papargyropoulou, E., Colenbrander, S., Sudmant, A.H., Gouldson, A., and Tin, L.C. The economic case for low carbon waste management in rapidly growing cities in the developing world: The case of Palembang, Indonesia. *J. Environ. Manage.* **163**: 11–9, 2015.

Stahel, W.R. Circular economy. *Nature*, **531**: 435–438, 2016.

Tatsioni, A., Vawa, E., and Ioannidis, J.P.A. Sources of funding for Nobel Prize-winning work: Public or private? *FASEB J.* **24**: 1335–1339, 2010.

Wadman, M. State of the donation. *Nature*, **447**: 248–250, 2007.

Chapter 5

Is Politics Part of the Cancer Research Affair?

5.1. Introduction

Niccolo Machiavelli's classic work, *The Prince*, written over 500 years ago, is a foundational text of political science, infamous for severing the link between politics and morality. The book is a guide to how people acquire and maintain power, exploring how societies work and how states can create the public spirit they need to survive. Chapter by chapter, *The Prince* guides a potential ruler through issues such as the best way to treat conquered states, how to plan a siege, and the right image to project to the masses. However, closer inspection reveals that this classic work is really a treatise on parochial altruism. Machiavelli understood that an individual's fate is inseparable from that of his or her group, and he had seen that weak states fall prey to those that are stronger and better organized. Therefore, his prescription for resisting invasion and suppressing plots focuses on building collective strength. The "friendship of the people", he says, is a ruler's best insurance against conspiracies, and a citizen militia is a more effective fighting force than foreign mercenaries. Even when Machiavelli advises that fear, and not love, is the safest guarantee of loyalty, he highlights the wider benefits of a fearsome reputation. Cruelty should be "as far as possible turned to the good of one's subjects". He writes that "by making an example or two" a ruler will prove more compassionate than those who, being too compassionate, allow disorderliness, which then leads to murder and mayhem.

The relevant concept of Machiavelli's book to our current cancer research is the need for an open dialogue between those who are making the policies and those who are doing the science. Whereas Machiavelli refers to friendship as an important liaison between the ruler and the people, here the analogy is that if we have a healthy dialogue between government and scientists, it will benefit the people in the long term. After all, the main function of a government is to ensure the safety and well-being of its citizens, in our context by fostering cancer research. This dialogue could be a good starting point for those seeking to understand the role of politics in cancer research, and it is probably task number one for the new cadre of cancer researchers. This is important because science-acquired knowledge is a long-term investment and as a consequence, it must not be at the mercy of market fluctuations or political whims, but must instead be a continuous flow over time. Only government can provide the stability necessary for the research to bear fruit. The problem is how to do it right.

5.2. The Science of Diplomacy

Machiavelli's "friendship of the people" concept says that amity with the people is a ruler's best insurance against conspiracies. Probably these words are not precisely right for our times but the translation of the concept into modern terms is what we call diplomacy — and the intricacy of politics in cancer research affairs certainly requires an understanding of diplomacy. US science diplomacy and international policy do not have a well-defined path for coexistence, but researchers with the right combination of tact, finesse, and negotiating skills may find niches in government agencies, science societies, nongovernmental organizations, or industry. Scientific acumen is not enough these days: to make their mark, scientists need not only a passion for international issues with scientific import, but also a knack for and interest in building relationships and a curiosity about other nations and cultures. This is what makes diplomacy work: when people that have the same goal but are coming from different positions are able to discuss a problem and move toward a solution.

Many European countries as well as the US government have long recognized the importance of science in international relations, diplomacy, and policy-making, but there aren't a huge number of scientists working for the government simply because the required skill set doesn't come easily to every scientist. Successful negotiating requires a knack for putting people at ease so that sensitive issues can comfortably be discussed. Therefore, our research scientists must build communication skills by presenting at conferences, participating in science workshops, learning how to write grant proposals, and most importantly learning how to listen effectively — treating one's counterparts from other cultures with respect is critical. A crucial quality scientists must possess is the plasticity to learn other disciplines. In summary, a scientist entering the diplomatic realm of politics must become familiar with the concerns and priorities of the diplomatic community because sharing objectives and working toward a common solution allow for the diplomacy in science to build bridges that can lead to local, national, and even global solutions.

5.3. The Democratic Process in Scientific Research

In recent years, academic scholarship and public discourse have become increasingly preoccupied with social and economic inequality, which has risen in many countries. This inequality means differences in three major domains: resources, research outcomes, and monetary or nonmonetary rewards. Scientific outputs and rewards are much more unequally distributed than other well-being outcomes, such as education, earnings, or health. As a result, a talented few can turn current successes into resources for future successes, accumulating advantages over time. Thus, it is accurate to say that although science rewards all participants, the most substantial rewards go to the top performers. In a certain way it is not different than a sporting event. In both pursuits after intense competition, the winners (or most successful participants) are rewarded with prizes, high visibility, a large contestant base, and the accumulation of advantages, including the diminishing of physical or cultural boundaries. This could be criticized as unfair, but this inequality in scientific rewards is based first on the principle that greater incentives for outstanding scientific work will ultimately result in greater benefits for humanity. An equally important factor is that scientists are supposed to be evaluated solely on merit, rather than on functionally irrelevant factors such as gender, race, nationality, age, religion, and class. For these reasons the apparent inequality in rewards is widely accepted and even favored (Cole, 2009).

Although there is general support for this concept of inequality in scientific research, there are, however, some concerns. One of them is that growth in high rewards in science has been limited, so it occurs much more slowly than the expansion of science itself. For example, it is well known that the number of Nobel Prizes is fixed; so, although many Nobel Prizes in natural science have been shared in recent decades, **the number of outstanding scientists who are not making the cut for this high scientific reward is significantly greater every year.** The future metric of these rewards is difficult to predict but if the trend continues they will lose their importance as a measure of scientific achievements, or worse they will be outliers in the system. The pyramid is getting bigger at the base and extremely narrow in the apex, thus the ideal of measuring success with this metric has less and less meaning.

Another inequality issue is the importance of the institutional environment to scientists. Research scientists affiliated with prestigious institutions are better rewarded than those who are not. This is because they are more productive in institutions that offer them a better infrastructure. Hence, greater institution-level inequality serves to intensify the individual-level inequality of research scientists. This can be better explained by saying that if a research scientist is hired by an institution that offers eventual tenure based on the researcher's progress that is an investment for the well-being of the institution because they can afford to allow the scientist to develop his or her full potential knowing they will both be rewarded at the end. But if the same research scientist is hired by an institution that is not economically viable and instead depends for its survival on the researcher bringing funding to the institution, it creates an inequality that will never stop. For example, in my institution, if you lose your funding, your tenure track allows you to stay for an additional 1–2 years but after that the institution is not able to sustain you any longer. Therefore in a system like this, the scientist's priority must be to continuously bring in external support for his or her scientific endeavor. Many of these institutions were created, as is discussed in another part of this book, with the magic thinking that government funds would continue to flow forever and that the institution would remain viable by the indirect benefit that each principal investigator would bring to the institution. However, this premise is easily broken in difficult economic situations, or when the numbers of applicants or institutions fighting for the same amount of resources get bigger and bigger. Although this is not the main cause of inequality, the reality is that if a research scientist's survival depends on maintaining the economic viability of the research institution, that in itself creates an inequality with those whose survival does not depend on the fluctuation of federal money or the financial system. Only those institutions that have created a safety net of endowment, which allows their scientists to continue their pursuits without financial worries, will be able to survive in times of duress, and create innovative and good science.

It is important to clarify here that permanent support of their research scientists by institutions does not directly translate to a bigger and better scientific output, but rather prestigious scientific institutions are able to maintain their reputation over time, probably due to their ability to recruit premiere scientists, whereas those institutions that are not making the mark are working with scientists who are risking their careers and gambling with their future resources. Some of these less privileged scientists are succeeding but many are not. As discussed in Chap. 1, at the end scientific endeavor is personal and success is individual; many succeed even in nonprestigious institutions and reach the apex of the pyramid. This human factor not only makes the system less unbalanced, but individual unpredictability also plays a critical role that cannot be quantified or generalized. This unpredictable human factor is what makes each research scientist unique because there is a driving force present in all those who are succeeding independently of the greatness of the sheltering institutions.

Finally, I would like to note two examples that illustrate the importance of the subject discussed. One is the work of Gregor Mendel in an isolated convent that took more than 30 years to be discovered but quickly became an integral part of scientific thought. The second is the history of Darwin and Wallace. Both produced an outstanding work, but Wallace was not in the mainstream whereas Darwin not only had the right environment but also the right network of friends and scientists who promoted his work, making it easily recognized and known; conversely it took almost a century to give Wallace the place in history that he deserves. The problem of inequality existed then and now, but there is hope the new information era could be a path to a solution (see Chap. 3).

5.4. Politicians Versus Scientists

We expect our politicians to have a realistic understanding of technology's promises and pitfalls and the ability to work comfortably with estimates and data that they use for statistical reasoning. We wish that politicians understood that the path toward good solutions is paved with uncertainty, trial, and error; that conclusions should be tentative; and that alternative views must be entertained. A more scientific thought process in politicians would make them more alert to the cognitive biases that can lead to irrational decisions. For example, politicians are often guilty of thinking in short-term solutions, i.e., the desire to enjoy rewards now rather than invest them for later. The government, on the other hand, is often involved in infrastructure projects or education programs that take several years to complete; most of the time, these projects outlast the careers of individual lawmakers. For us scientists, we are comfortable thinking about processes on different timescales — millions or even billions of years. Scientific thinking can thus build strong arguments for investment in roads, bridges, trains, and laboratories that will not produce profits tomorrow but will pay off powerfully in the decades to come. This difference in the thought process makes many lawmakers uncomfortable with scientific reasoning. Uncertainties are part of the life of scientists, whereas nonscientists are less comfortable with this doubt. For example, a failure to understand ordinary fluctuations in noisy climate data allows some members of Congress to believe that claims of human-induced climate change are a hoax, or that the data are so chaotic no policy action can be devised.

So how can we increase scientific thinking in our governmental branches? One way, of course, is to elect more scientists by strongly encouraging scientists to seek political office. But that is an unlikely solution as most scientists prefer to stick with their chosen fields. More viable is the idea that nonscientists could learn to feel comfortable thinking like scientists. This is not without precedent — not all legislators hold law degrees, but all must be comfortable thinking like lawyers when drafting a bill or reading a statute. Thus when legislators need to think like lawyers, they turn to legal professionals for help, and the same kind of guidance should be offered to legislators by scientists and, hopefully over time, they would pick up more scientific ways of thinking.

But to expect that politicians will think scientifically is probably too much to ask; however, it should be considered that a better partnership between science and government could be healthy and beneficial for all of us. This implies also that scientists would be educated to understand the science of political and social problems behind legislative decisions. In the same way that scientists have often been considered as outlier and extravagant individuals, politicians have been associated with corruption and selfishness and so on, but the reality is that neither the scientists nor the politicians are simply clusterings of adjectives. In my perception, scientists are driven by an internal force that makes them curious to find the cause or remedy to some specific problem of nature, whereas politicians, on the other hand, are driven by the ideal of bringing solutions to people. But both groups need to endure a long process to be successful; therefore, to find an adequate and productive partnership between these two types of individuals is one of the functions of the new cadre of researchers, mainly those that are working in cancer, which affects so many people.

5.5. The Inequality Problem as the Basis of Political Action in Cancer Research

The protesting of the 2011 Occupy movements has faded but an international conversation about inequality has arisen and continues to echo around the world. Today the top 1% of the population controls nearly 20% of US income, up from about 8% in the 1970s. But inequality is increasing within the 99%, too, as a consequence of a growing premium on college and postgraduate education. According to surveys by the Census Bureau, in 2012 the richest 20% of Americans enjoyed more than 50% of the nation's total income, up from 43% in 1967. The middle 20% — i.e., the actual middle class — receives only about 14% of all income, and the poorest get a mere 3%. Many nations, especially emerging economies, have even larger gaps between the superrich and the poor (Piketty and Saez, 2014).

Countless studies have established the harmful effects of disadvantageous circumstances on education and health. These influences can begin early in life, even prenatally. Researchers in this field are still exploring whether the stress of being low-ranked itself adds to the poor's burden, possibly causing illness and even early death. In addition, psychological mechanisms may spur a negative feedback loop in which poor individuals behave in ways that help keep them poor. Harsh as life can be for those at the bottom, the opportunity to move up the ladder can compensate. Newly available data from taxes and other records promise to yield insights into intergenerational mobility, in which children advance from their parents' socioeconomic status. But so far, there is a relatively limited view of how and why people move into different social and economic classes.

5.5.1. *The ancient roots of inequality*

Economic disparities were common in the Roman Empire, where 1.5% of the empire's households controlled 20% of the income by the late second century CE. The earliest elites emerged about 10,500 years ago, when people successfully domesticated plants and animals and settled in large permanent villages. In this view, agriculture led to the production of surpluses and then to the emergence of managers, craftspeople, and other specialists who eventually gained control over the extra resources. **Basically it was the ownership of small, resource-rich areas, and the ease of bestowing them on descendants, that fostered inequality, rather than agriculture itself; however, agricultural life forced the management of the land.** Inequality markers became more common between 10,500 and 8,200 years ago, as early farmers began sowing domesticated einkorn wheat and other plants and tending domesticated sheep and goats. But signals of incipient inequality appeared well before that, between 14,500 and 12,800 years ago with the Natufian culture. The Natufians inhabited a region in the eastern Mediterranean. It was a sedentary or semisedentary society, and it is believed that the Natufian communities are possibly the ancestors of the builders of the first Neolithic settlements of the region. There is archeological and physical anthropological evidence for a relationship between the modern Semitic-speaking populations of the Levant and the Natufians (Bengtson, 2008). In Natufian society the wealthiest 8%, for example, were decorated with pendants or marine shells such as *Dentalium*, imported or traded from as far as 400 km away. Natufians also placed carved artworks in a few graves, built houses of varying sizes, and produced large goblet-shaped stone mortars well-suited for preparing or serving food at feasts (Pringle, 2014). Natufians apparently "were harvesting wild plants in large quantities and storing cereal grains as well and these stored surpluses of wild cereals may have given some Natufian hunter-gatherers an edge over others. Those surpluses could allow people to begin manipulating things, giving away food and so establishing some dominance behaviors" (Pringle, 2014).

Owning land, while important in its own right, is also the basis for inheritance, another important factor in establishing inequalities. Only material forms of wealth, such as land and livestock valued by farmers, were readily handed down to children. This control of wealth was not found in the egalitarian lifestyle of hunting and gathering societies. But once farmers owned the land, they could control access to their fields, protect them, and leave them to their heirs (Pringle, 2014). Resource concentration is a key factor in explaining inequality among both farmers and ancient salmon fishers. According to Pennisi (2014), **"Farming vastly increased the productivity of small patches of land and a small number of animals".** People who owned particularly fertile patches of farmland had a good opportunity to not only become wealthy themselves but also to pass on that wealth, in part because the land was defendable against others. As agricultural societies developed, so did more elaborate hierarchies, evolving into hereditary chiefdoms and eventually kingdoms. In these increasingly complex societies, chiefs and kings came up with new strategies for amassing surpluses and concentrating wealth and power. Many chiefs created economic bottlenecks in trade routes, then collected payments from merchants for safe passage and used the surplus to finance specialized warriors to defend and extend their rule. Material culture also became ever more sophisticated, multiplying into innumerable kinds of highly concentrated and easily transmitted forms of wealth, from copper ingots to gold jewelry. All of these developments led to ever greater levels of inequality.

While it is difficult to forecast the disappearance of inequalities, an interesting sign is that in the present time societies are moving toward knowledge-based economies; wealth increasingly reflects know-how, social skills, and networking — factors that cannot be transmitted across generations as easily as plots of land or stock portfolios. Therefore there is a possibility of movement to a more egalitarian society or at least a reduction in the gap between those that have and those that do not have.

5.5.2. *Supporting data on inequality*

A huge database of tax records that Piketty and a team of 30 researchers around the globe assembled from more than 20 countries, including the United States, challenges the long accepted idea that wealth and income would be more evenly distributed within nations as they developed and suggests that even the best run capitalist economies concentrate riches at the top. The reason? In the long run, he says, the return paid to owners of capital is higher than the rate of economic growth (Marshall, 2014). Piketty unearthed French data on wealth going back to the 1789 French Revolution, as well as a century's worth of income tax data that hadn't been analyzed systematically. Many such records have been ignored, Piketty (2014) claims, writing, "The historical and statistical study of tax records falls into a sort of academic no-man's-land, too historical for economists and too economic for historians". Piketty combines income, inheritance, and national wealth data to reach a striking conclusion: Capitalism concentrates riches at the top of society. Piketty argues this is because the rate of return on capital (labeled r) is higher than the overall rate of economic growth (labeled g) over the long run. This simple formula ($r > g$) means that families who own capital tend to acquire more and more wealth. Piketty says that inequality in the developed economies and particularly in the United States has reached an "extreme" point that could lead to "terrifying" disparities in the future and threaten democracy. This inequality also has been explained by the analogy of Cho (2014) about the concept of entropy, a measure of disorder in a physical system such as a gas. Just as a gas evolves to a state of maximum entropy, random churning in the economy ensures that the income distribution naturally tends to this inequitable form. Cho's argument suggests that although social and economic policy can help individuals edge ahead or perhaps boost everybody's fortunes, nothing short of radical intervention can overcome the forces of randomness and transform the lopsided distribution.

5.5.3. *The consequences of inequality*

Compared with the wealthy, poor people are less healthy. A child born in Norway can expect to live roughly 30 years longer than one born in Afghanistan. In the United States, on average, people in the highest income group can expect to outlive those in the lowest income group by more than 6 years. Preventable illnesses caused by poor nutrition and lack of education and health care account for much of the disparity. Investing in health care and making it widely available can boost the health of those at the bottom (Underwood, 2014). According to Underwood (2014), redistributing wealth to the lower end of the curve helps, too. But to understand the issue, we must first understand *how* being low on the social ladder affects health. The next question is whether a society's health is worse when the rungs are far apart. The issue is a hot topic just below the surface in policy debates and has erupted recently in impassioned editorials. Some argue, paraphrasing Roman philosopher Seneca the Younger, that "to be poor in a wealthy society is the worst kind of poverty". But will it send you to an early grave?

When a population moves beyond abject poverty, rank in the social hierarchy, not income, ultimately determines how healthy people are. From the dangerous streets of Chicago's South Side to the neatly tended homes of a Helsinki suburb, the link between low status and poor health has now been found in many different countries and contexts. More controversial is whether overall population health is worse in more unequal societies. Kondo *et al.* (2009) published a meta-analysis of epidemiological studies linking inequality and health in about 60 million people around the world. They found an excess mortality risk of 8% for every 0.05 unit increase in a country's Gini coefficient, the most commonly used statistical measure of the gap between rich and the poor. Although such an effect may seem modest, when extrapolated to the global population it suggests that leveling income inequality could help avert more than 1.5 million deaths per year worldwide — assuming the effect is causal, Kondo cautions.

Inequality breaks down social values, such as trust and support, that protect against both physical and mental illnesses. In an article published in *The New York Times*, epidemiologists Wilkinson and Pickett (2009) of the University of York in the United Kingdom take the argument even further. They claim that the reason more unequal countries like the United States see higher rates of schizophrenia and other mental illnesses is because inequality causes "social corrosion" that damages the individual psyche.

The problem of inequality has been tackled in many different ways throughout history. For example, "In China Mao Zedong waged war on inequality of all kinds. The administration seized property from privileged classes, imprisoned intellectuals, and appointed teams of workers to run universities. The revolution upended the class structure, and the party campaigned against inherited wealth and gender discrimination. By the time the Cultural Revolution ended and Mao died in 1976, the government had mandated a bland unisex style of dress and effectively abolished property ownership. Society had ostensibly been 'leveled off': even if in practice the new system concentrated resources in the hands of party cadres" (Hvistendahl, 2014). When this system ended in 1980 the egalitarian society had been transformed into a society with high levels of inequality; the richest 10% now make 13 times as much as the poorest 10%, compared with five times as much in the United States (Hvistendahl, 2014).

An indicator for India was the Gini coefficient, a common index of income inequality ranging from 0, in which everyone makes the same income, to 1, in which a single rich person would get a country's entire income. Government surveys based on expenditures and excluding income data had found figures in the 0.30s — below the 0.40 level in the United States. In 2010, the Indian survey found a Gini coefficient of 0.52 — close to China's, which scholars most recently estimated at 0.55. At a time when attention is focused on inequality in the developed world, that's a sharp reminder that while the worst inequalities are often in emerging economies, **inequality in high-income countries "still falls well below levels found in low- and middle-income countries.** The middle classes in China or India are" still rather poor within global comparisons. However, the Chinese appear remarkably tolerant of income gaps. Asked why people are poor, 61% said a lack of ability was an important cause, far higher than in any other country" (Hvistendahl, 2014).

According to Mervin (2014), when studying disparity or how to measure it, it is important to make a distinction between two different ideas: economic inequalities and social, or intergenerational, mobility. The former is the gap in wealth between those in the penthouse and the poorhouse, which data show has reached high levels in the United States and is growing around the world. The latter is whether children, as adults, "get at least as far as their old man got" on a ladder defined, not only by income but also by social factors including education, occupation, and where you live (Mervin, 2014). Unfortunately the data are not clear enough to make a distinction that goes to the root of the problem.

In a report, Piketty and Saez (2014) described basic facts regarding the long-run evolution of income and wealth inequality in Europe and the United States. Income and wealth inequality were very high a century ago, particularly in Europe, but dropped dramatically in the first half of the 20th century. The long-run dynamics of income inequality is the most difficult part to study because income inequality combines forces arising from the inequality of capital ownership and capital income as well as other forces related to the inequality of labor income. Income is a flow. It corresponds to the quantity of goods and services produced and distributed each year. Income can be defined as the sum of labor income (wages, salaries, bonuses, earnings from nonwage labor, and other remuneration for labor services) and capital income (rental income, dividends, interest, business profits, capital gains, royalties, and other income derived from owning capital assets).

According to Card *et al.* (2001), the economic model most widely used to understand inequality is based on the idea of a race between education and technology. That is, the expansion of education leads to a rise in the supply of skills, while technological change leads to a rise in the demand for skills. Depending on which process occurs faster, the inequality of labor income will either fall or rise. One proposed explanation for the increase of inequality in recent decades has been the rise in the global competition for skills, itself driven by globalization, skill-biased technical change, and the rise of information technologies. However, such skill-biased technological progress is not sufficient to explain important variations between countries: The rise of labor income inequality was relatively limited in Europe and Japan compared to the United States, despite similar technological changes. In the very long run, European labor income inequality appears to be relatively stable. This suggests that supply and demand for skills have increased at approximately the same pace in Europe. It is possible that the large increase in United States labor income inequality in recent decades can be explained by insufficient educational investment for large segments of the US labor force. If this is the case, massive investment in higher education would be the right policy to curb rising income inequality (Card *et al.*, 2004). Yet, although this view is very appealing, it does not account for all of the facts. In particular, the race between education and technology fails to explain the unprecedented rise of labor incomes at the very top that has occurred in the United States over the past few decades. A very large part of the rise in the top 10% income share comes from the top 1% (or even the top 0.1%). This is primarily due to the rise of top executive compensation in large financial and nonfinancial US corporations. Therefore, inequality does not follow a deterministic process; there are powerful forces pushing alternately in the direction of rising or shrinking inequality. Which one dominates depends on the institutions and policies that societies choose to embrace (Piketty and Saez, 2014).

A 2014 work published by Autor (2014) in *Science* considers the role of the rising skill premium in the evolution of earnings inequality. There are three reasons to focus a discussion of rising inequality on the economic payoff to skills and education. First, the earnings premium for education has risen across a large number of advanced countries in recent decades, and this rise contributes substantially to the net growth of earnings inequality. In the United States, e.g., about two-thirds of the overall rise of earnings dispersion between 1980 and 2005 is roughly accounted for by the increased premium associated with schooling in general and postsecondary education in particular (Goldin and Katz, 2008). Second, despite a lack of consensus among economists regarding the primary causes of the rise of very top incomes, an influential literature finds that the interplay between the supply and demand for skills provides substantial insight into why the skill premium has risen and fallen over time. A third reason for focusing on the skill premium is that it offers broad insight into the evolution of inequality within a market economy, highlighting the social value of inequality alongside its potential social costs and illuminating the constructive role for public policy in maximizing the benefits and minimizing the costs of inequality (Alvaredo *et al.*, 2009; Bivens and Mishel, 2013; Bonica *et al.*, 2013; Goldin and Katz, 2008).

Workers' earnings in a market economy depend fundamentally on their productivity, i.e., the value they produce through their labor and in turn, workers' productivity depends on two factors: One is their capabilities, concretely, the tasks they can accomplish, and second is their scarcity.

The fewer workers available to accomplish a task and the more employers need that task accomplished, the higher the economic value of the workers is relative to that task. In conventional terms, the skill premium depends upon what skills employers require (skill demand) and what skills workers possess (skill supply). A technologically advanced economy requires a literate, numerate, and technically and scientifically trained workforce to develop ideas, manage complex organizations, deliver health-care services, provide financing and insurance, administer government services, and operate critical services (Alvaredo *et al.*, 2009; Bivens and Mishel, 2013; Bonica *et al.*, 2013; Goldin and Katz, 2008).

Goos *et al.* (2010) made an interesting analysis that is summarized as follows:

> the past three decades of computerization, in particular, have extended the reach of this process by displacing workers from performing routine, codifiable cognitive tasks (e.g., bookkeeping, clerical work, and repetitive production tasks) that are now readily scripted with computer software and performed by inexpensive digital machines. This ongoing process of machine substitution for routine human labor complements educated workers who excel in abstract tasks that harness problem-solving ability, intuition, creativity, and persuasion — tasks that are at present difficult to automate but essential to perform. Simultaneously, it devalues the skills of workers, typically those without postsecondary education, who compete most directly with machinery in performing routine-intensive activities. The net effect of these forces is to further raise the demand for formal education, technical expertise, and cognitive ability.

The persistently rising demand for educated labor in advanced economies was first noted by the Nobel Prize–winning economist Tinbergen (1974) and is often referred to as the "education race" model (Goldin and Katz, 2008). Its primary implication is that if the supply of educated labor does not keep pace with persistent outward shifts in demand for skills, the skill premium will rise. In the words of the Red Queen in Lewis Carroll's *Alice's Adventures in Wonderland*, "…it takes all the running you can do, to keep in the same place." Thus, when the rising supply of educated labor began to slacken in the early 1980s, a logical economic consequence was an increase in the college skill premium. To more formally account for the impact of the fluctuating growth rate of supply of college-educated workers on the college wage differential (Goldin and Katz, 2008).

Of course, this set of facts raises another puzzle: If slackening college supply sparked rising inequality, what caused rising US postsecondary achievement to grind to a sudden halt in 1982? Work by Card and Lemieux points to the United States' involvement in the Vietnam War as the critically important factor.

> Because draft-eligible males in the Vietnam era were often able to defer their military service by enrolling in postsecondary schooling, the war artificially boosted college attendance. This created something of a glut of college enrollments in the late 1960s and early 1970s, which in turn depressed the college earnings premium in the 1970s and likely reduced the attractiveness of college-going absent the military draft. Thus, when the war ended in the early 1970s, college enrollment rates dropped sharply, particularly among males. The fall in enrollment produced a corresponding decline in college completions half a decade later, and a surge of inequality followed.

But this supply–demand explanation for the rise of US inequality may appear almost too simple to be credible. After all, it was the comparison of just two economic variables: the college wage premium and the supply of college graduates in the US workforce. But a host of rigorous studies commencing with Katz and Murphy (1992) confirm the remarkable explanatory power of this simple supply–demand framework for explaining trends in the college versus high school earnings gap over the course of nine decades of US history, as well as across other industrialized economies, most notably the United Kingdom and Canada, and among age and education groups within countries (Goldin and Katz, 2008; Katz and Murphy, 1992; Card, 2001; Crivellaro, 2014). The United States was far from the only Western country to experience this surge.

5.6. Conclusions

We have discussed several important issues such as diplomacy, how the democratic process influences scientific research (mainly cancer research), and how the interaction between politicians is necessary for moving the agenda of cancer research ahead. Finally I have dedicated a great portion of the chapter to the issue of inequality, making reference to publications that have extensively discussed this issue. The main point is that although we as cancer researchers are extremely focused on our own endeavors and working toward a meaningful end, there are political issues that directly or indirectly affect our lives and therefore it is important that we are cognizant of them so that we might have a better grasp of our daily reality.

References

Alvaredo, F., Atkinson, A.B., Piketty, T., and Saez, E. The top 1 percent in international and historical perspective. *J. Econ. Perspect.* **27**: 3–20, 2013.

Autor, D.H. Skills, education and the rise of earnings inequality among the "other 99 percent." *Science*, **344**: 843–851, 2014.

Bengtson, J.D. (editor). *In Hot Pursuit of Language in Prehistory: Essays in the four fields of anthropology.* In honor of Harold Crane Fleming. Philadelphia: John Benjamins Publishing, 2008.

Bivens, J., and Mishel, L.J. The pay of corporate executives and financial professionals as evidence of rents in top 1 percent incomes. *J. Econ. Perspect.* **27**: 57–78, 2013.

Bonica, A., McCarty, N., Poole, K.T., and Rosenthal, H. Why hasn't democracy slowed rising inequality? *J. Econ. Perspect.* **27**: 103–124, 2013.

Card, D., Lemieux, T. Going to college to avoid the draft: The unintended legacy of the Vietnam War. *Am. Econ. Rev.* **91**: 97–102, 2001a.

Card, D. and Lemieux, T. Can falling supply explain the rising return to college for younger men? A cohort-based analysis. *Q. J. Econ.* **116**: 705–746, 2001b.

Card, D., Lemieux, T., and Riddell, W. Unions and wage inequality. *J. Labor Res.* **25**: 519–559, 2004.

Cho, A. Physicists say it is simple. *Science*, **344**: 828, 2014.

Cole, J.R. *The Great American University: Its Rise to Preeminence, its Indispensable National Role. Why it Must Be Protected.* New York: Public Affairs, 2009.

Crivellaro, E. College wage premium over time: Trends in Europe in the last 15 years. University Ca' Foscari of Venice, Department of Economics Research Paper Series no. 031WP/, 2014.

Goldin, C.D. and Katz, L.F. *The Race between Education and Technology.* Cambridge: Harvard University Press, 2008.

Goos, M., Manning, A., and Salomons, A. Explaining job polarization in Europe: The roles of technology, globalization and institutions. CEP Discussion Paper No. 1026 ISSN 2042-2695, 2010.

Hvistendahl, M. While emerging economies boon equality goes bust. *Science*, **344**: 832–835, 2014.

Katz, L.F. and Murphy, K.M. Changes in relative wages, 1963–1987: Supply and demand factors. *Q. J. Econ.* **107**: 35–78, 1992.

Kondo, N., Sembajwe, G., Kawachi, I., Van Dam, R.M., Subramanian, S.V., and Yamagata, Z. Income inequality, mortality, and self rated health: Meta-analysis of multilevel studies. *BMJ.* **339**: b4471, 2009.

Marshall, E. Tax man's gloomy message: The rich will get richer. *Science*, **344**, 826–827, 2014.

Mervin, J. Tracking who climbs up and who falls down the ladder. *Science*, **344**: 836–837, 2014.

Piketty, T. and Saez, E. Inequality in the long run. *Science*, **344**: 838–342, 2014.

Pringle, H. The ancient roots of the 1%. *Science*, **344**: 822–825, 2014.

Pennisi, E. Our egalitarian Eden. *Science*, **344**: 824–825, 2014.

Tinbergen, J. Substitution of graduates by other labour. *Kyklos*. **27**: 217–226, 1974.

Underwood, E. Can disparity be deadly? *Science*, **344**: 829–831, 2014.

Wilkinson, R., Pickett, K. *The Spirit Level: Why More Equal Societies Almost Always Do Better*. London: Allen Lane, 2009.

Further Reading

Allison, P.D. and Long, J.S. Departmental effects on scientific productivity. *Am. Sociol. Rev.* **55**: 469–478, 1990.

Cole, J.R., Cole, S. *Social Stratification in Science*. Chicago: University Chicago Press, 1973.

Ehrenberg, R.G., Bognanno, M.L.J. Experimental comparison of multi-stage and one-stage contests. *J. Polit. Econ.* **98**: 1307–1324, 1990.

Frank, R.H. and Cook, P.J. *The Winner-Take-all Society*. New York: Penguin, 1995.

Katz, L.F. and Autor, D.H. Changes in the wage structure and earnings inequality. In *Handbook of Labor Economics*, D. Card, O. Ashenfelter (eds.). Vol. 3, pp. 1463–1555. Amsterdam: Elsevier-North Holland, 1999.

Lazear, E.P. and Rosen, S. Rank-order tournaments as optimum labor contracts. *J. Polit. Econ.* **89**: 841–864, 1981.

Merton, R.K. *The Sociology of Science. Theoretical and Empirical Investigations*. pp. 267–278. Chicago: University Chicago Press, 1973.

Price, D.J. *Little Science, Big Science*. New York: Columbia University Press, 1963.

Xie, Y. and Killewald, A.A. *Is American Science in Decline?* Cambridge: Harvard University Press, 2012.

The Role of Society in the Training of Cancer Researchers

6.1. Introduction

National investment in research is always an important issue in both developed and developing countries. However the role of society in supporting research varies greatly. For example, consider Centre Nationnal de la Recherche Scientifique (CNRS) and Institut national de la santé et de la recherche médicale (Inserm), which are representative of France's research endeavors. Just after World War II, France created these agencies, separate from the universities, to do basic research. Not long after, Argentina created the National Scientific and Technical Research Council (CONICET), similar to the INSERM in France. At first such a setup was found only in Communist countries, in particular the USSR (Union of Soviet Socialist Republics) and China. Now, these countries have abandoned this model but it still persists in France and Argentina. In the United States, the vast majority of research operators are universities. Research organizations are expected to behave more like real funding agencies, serving the universities and research institutions. The basic rule is that institutions that are running research, mainly cancer research, must be led by the same principles that guide the society or country where they are located. A society that does not think of the future could still produce outstanding institutions in which research is matured and run by exceptional individuals.

6.2. Society as a Driving Force in Cancer Research

The main question is why certain nations or societies are more successful than others in their research endeavors. Many explanations can be given,

111

such as abundance of natural resources, absence of serious conflicts, or specific historic and geographic advantages. Although these are all reasonable explanations, a primary determinant of success is how the leaders of major institutions, governmental and nongovernmental, are selected; how they in turn choose their deputies; and under what incentives they must operate. Definitely the society must select leaders with a focus on long-term agendas. In the case of cancer research, many shortcut approaches have been used, and the reality is that none of them have been effective enough. Therefore, the society needs to understand that science requires time to mature the knowledge and understand process, and this is particularly true in cancer research. The important concept is that society must use not only a democratic but also a merit-based system to choose leaders who are able to solve real problems. The election of outstanding individuals to prominent positions provides leaders with the knowledge and self-confidence to, in turn, choose the right people, those in whom trust must be deposited. As a rule poorly qualified individuals select unqualified individuals. In a 2013 *Science* article Bruce Alberts writes, "his advice is to managers to hiring people who, at least in some respects, appear to be more talented than you are. This requires honest humility. But only in this way can a leader hope to achieve his or her goals. It is only through a meritocracy in which leaders encourage creativity from outstanding subordinates and are primarily rewarded for long-term, rather than short-term achievement" (Alberts, 2013). This goal is achievable, but depends on the societal environment in which the individuals live. An article published in *Nature* by Gachter and Schultz (2016) shows that a country's prevalence of rule violations, which for this study included tax evasion, corruption, and political fraud, is positively associated with the tendency for residents of that country to lie for small amounts of extra cash. The study was performed only in a small number of people, 2568, from 23 countries. Among the conclusions of this study are that low exposure to rule violations increases people's intrinsic honesty, not vice versa; that participants from more corrupt countries lied more than those from less corrupt ones; and that when people interact with a lying partner, they are likely to lie as well (Gachter and Schultz, 2016). This publication does not explain the complex process of corruption and human behavior but does hint that the moral compass of society is determined by human beings, further implying that honesty is important as is the capacity to understand long-term consequences of small corrupted acts. With honest and qualified people in positions of power and better-described decision-making in human health, it is possible that the beginning of an epidemic could be determined by the emergence of few outliers, and thus quickly contained, assuming adequate people are in place to spot them. As I have described in Chap. 4, the contamination of water with lead was a mishandling of information and resources from people that were in charge of the public health. Cancer researchers must institute adequate preventive strategies, in addition to adequate diagnostic procedures and treatment, to provide the largest benefit to society.

6.3. Society Must be a Guardian of the Integrity of Research Support

It has been indicated by Suresh (2012), in his article in *Science*, "that the scientific research enterprise in this expanding global endeavor requires an open flow of information with transparent processes to promote rigorous peer review and scientific integrity. For this aim to be accomplished it requires a social understanding of what is expected from science" and furthermore from cancer researchers. "Among the criteria that need to be held in this interactive society are that the science provides an expert assessment of the problem under study, that the process of evaluation of the scientific data be transparent and impartial with adequate confidentiality, and that there is ethical integrity in its evaluation and dissemination". Although these basic principles are already in place in developed economies, it is not completely settled in those countries with developing economies or in areas where the research endeavor is not a national priority. It is not the objective of this section to parse details about specific research activity in different countries but to provide instead an overview of the general principles that are needed for science to be accepted as a social responsibility.

The globalization of human endeavor in the economy must foster integration not only at the level of goods consumption but also in how scientific research is conducted and shared. Respect for intellectual property and also patenting could be good starting points as society defines its role in cancer research. A practical problem in this idyllic global economy is international collaboration in performing clinical studies or clinical trials. This requires building a collaborative group of experts whose skills complement those of the principal investigator, who originates the idea. In addition to creating the right team, it also requires careful planning, dealing with regulations, and adapting to unexpected circumstances in a different culture (Jung, 2008).

An article published in *Science* by Nurse *et al.* (2013) discusses how to build better research institutions. The pledge is that "biomedical research has to integrate biological, nonbiological, and clinical disciplines, and its application requires interactions with hospital and commercial partners. This can be facilitated by research institutions with an environment that supports strong interdisciplinary interactions between scientists: a place where laboratory biologists are encouraged to collaborate with clinical researchers to understand the medical implications of their work, with pharmaceutical companies for the translation of discoveries into treatments, and with physical scientists to expand their thinking and repertoire of experimental approaches". This type of institution already exists, but the problem is that some of them have a brilliant and meteoric start and then funding gets in the way of maintaining the needed research environment. Most of these dream institutions are created around a Nobel Prize winner or with seed money from a wealthy donor or by a government spur of funding driven by a visionary in the public health system. Whatever the initial driving force is, in time these well-conceived institutions run out of steam and enter into the pool of institutions already competing for the same limited public funds. This kind of cyclicity of the research institutions occurs over and over again. Therefore, as a society we must recognize that something is not right in the way that we conceive and maintain these dream institutions. It is the responsibility of society to preserve those institutions that show themselves to be germinal centers of creativity and innovation. This must be a priority to society.

Another important responsibility that society, through its universities and research institutions, must assume is the well-being of its scientists. It is not enough to provide positions that afford scientists a good standard of living and the ability to face the challenges of publishing, acquiring funding, and balancing the demands of work and personal life, but society must also provide them security, namely tenure track positions. The new cadre of cancer researchers are facing fierce competition, not only for grants in a situation where the overall funding dollar is decreasing, but also for increasingly scarce tenure track positions, which are not increasing with the demand for them. History tells us that researchers, mainly those in academia, are used to economic constraints but it is the global economy that is infringing upon the personal growth of the research scientists. It is on this point that society as a whole and in each country, irrespective of whether it has a developed or developing economy, must create a framework or structure for saving the scientific endeavor. It is an investment in human resources that cannot be allowed to depend on economic fluctuation of the market. It is a political and social responsibility to provide support for the scientists.

6.4. Social Diversity in Cancer Research

In a multicultural society like the United States, it is important to encourage participation of all members of society, especially women and those that are considered minorities. In recruiting women and minorities it is important to discuss the relevance of research in science, especially in cancer research. The emphasis that needs to be made is that a person who decides to be a cancer researcher must be able to reach their full potential pursuing that endeavor. Every time that I interview a candidate, for either a research or an internship position, I try to determine the individual endowment that the person has to succeed as a scientist. The inner qualities must emerge beyond the physical appearance provided by gender and race.

It is true that we as a society must also strive to avoid disparities; we must also not allow differences in salaries or advancement opportunities between females and males. It has been reported that there is a glass ceiling for women and minorities despite the great advances that we have seen in the last 20 years. An article published by Yang (2011) discusses the finding that minority professionals have a lower chance than white professionals of reaching higher level executive positions, and Asian Americans have the lowest chance for success compared to blacks and Hispanics. Although there is not any rational explanation for this disparity, it is clear that if we as a society want to maximize all the intellectual power that we possess for innovation we must consider and eliminate biases rooted in our behavior. As indicated by Huang (2011), this probably can be achieved by: "1–developing a complete intolerance for the casual discrediting or minimizing of contributions and accomplishments made by women and minorities; 2–making sure that the bar is not set higher for women and minorities than for white males; 3–removing any inequity in pay; 4–promoting qualified minorities and women to prestigious leadership positions in a timely way, and 5–by providing committed leadership for accomplishing these goals". The new cadre of cancer researchers must bear in mind these principles.

6.5. Social Clues

I would like to discuss few of the many concepts that we must have present all the time in the process of training cancer researchers.

6.5.1. *Transdisciplinary sciences*

Research scientists succeed when innovative thinking is introduced in the equation; this means that it is important to master not only the methods of science but also the ability to be open to other levels of consciousness, such as art. This, not a new concept and the intellectual tradition of the science of today, has emerged in conjunction with art. It is not doubted that the collaboration of art and science forces our brains to engage with different ideas and problems. The same applies to bridging social or other humanitarian disciplines, such as philosophy and logic, that are spheres of knowledge, which can span our creative and innovative vision of the problem to solve. However the most basic difficulty in exploring and bridging disciplines is not the concept or the scientists but finding the funding opportunities to do it. However, this concept of transdisciplinary science is gaining momentum, especially in areas of health and environment, where the social impact is great. Therefore, the next generation of cancer researchers must learn to read and understand the social clues provided by those disciplines beyond the horizon and expertise of the biological sciences.

6.5.2. *Communication as part of our social clues*

Our human behavior in society has evolved over 21 centuries of knowledge, influenced by our exposure to different cultures, beliefs, tastes, and expressions of beauty. Independently our positions have changed on homosexuality, the rights of women and racial minorities, slavery, child labor, and the abuse of animals, to cite a few issues; we are now more attentive to and perceptive of different positions regarding these issues. The fact that we are openly discussing these issues has created a different way to see the problem. The basic principle is that when we know and share our ability to communicate with other human beings it brings closeness, even affection. The possibility of easily connecting through social media or even better by traveling helps us understand our economic interdependence and widens our social circles. This is not to say that our views on each of the issues mentioned earlier instantly change when we come into contact or communication with other people; rather it is a way to open our minds to the perspectives of others. That is a social clue that we start learning in childhood with the appropriate influence of parents, and later reaffirm with the social networks that we start to develop, both personally and professionally. What we are learning as a society is that better understanding of others begins with language. In my opinion, language is the major barrier among humans. Only when the language barrier is broken do we start to understand that the feelings and ideas of others could be very similar to our own, and we become more open to different viewpoints because better communication allows better understanding. If humans are refractory to communication as a basic social clue, there is not any way that we as researchers can pursue our ideas further. When I went to Japan for the first time in the 1970s, very few scientists spoke English but they understood the challenge; 20 years later English was the norm in scientific meetings and it was possible to talk with most of the people that were working in a research laboratory.

We as cancer researchers must be the first ones to understand the social clues of isolation created by barriers and radical ideas. Breaking the barriers of language allows us to easily approach others and use the natural ability of every human being to be a storyteller, sharing our adventures and interests. The influence that we could have on others is basically limited only by our ability to communicate and our willingness to break the isolation of a single human being.

6.5.3. *Scientists as storytellers, a way to connect with society*

What makes a great book unique is when its story has the power to change the way that people think. Examples of these thought-provoking books are *Uncle Tom's Cabin, To Kill a Mockingbird,* and *Animal Farm.*

Uncle Tom's Cabin, written in 1852 by Harriet Beecher Stowe, depicts a grim picture of life under slavery. Uncle Tom, the main character of the book, is eventually beaten to death by his overseer. The book helps readers to not only understand slavery but also the beginning of its abolition. *To Kill a Mockingbird* is a novel by Harper Lee published in 1960 that deals with important issues such as rape, racial inequality, and discrimination. The novel delves into not just these incendiary issues, but also their social implications and how a human being like Atticus Finch can change the perceptions of the society in which he lived. George Orwell's *Animal Farm,* written as an allegory of the 1917 Russian Revolution, remains just as applicable to today's rebellions against dictators around the world.

The list of powerful books like these is endless; however, the point is clear: we as cancer researchers need to learn how to tell our stories so that people understand not only what we want to do with our research but also how we hope to educate them in the prevention and treatment of the disease.

Literature writers have long used scientific knowledge to make their points, such as in George Eliot's *Middlemarch.* In this novel, Eliot not only makes reference to Andreas Vesalius and Bichat's theory of primary tissues like webs, but also uses these concepts in the structure of the story. This illustrates the need for cancer researchers to use literary means that are part of the cultural background of the cancer sufferer to tell the story. The use of analogies like the similarity of the spiral in the snail shell to the helix of the DNA structure can really help people visualize the individual DNA helix structure and advance the concept of personalized medicine, for example.

The scientific mind works with logic and facts but the reality is that imagination and fantasy play an important role in society, supported by a great cultural literary background displayed in the movies, TV shows, advertisements, and many other influences in our ordinary life. Therefore, it is the task for the new cadre of cancer researchers to fill this gap or to make these two different poles a more coherent one. We scientists must learn from literature writers to tell our stories in the same way that Dostoyevsky and Tolstoy explained the convolution of the thinking mind to the readers.

6.6. Debt to Society or a Society in Debt

Student debt is one of the greatest problems for the millennial generation. According to Goldman Sachs, for mentioning a source, it indicates that this generation can be characterized to be between 18 and 35 years of age and by being socially liberal, racially diverse, digitally savvy, economically distressed, and having a collective debt of *1 trillion* dollars. The millennials have a high level of unemployment, 4 out of 10 are living with their parents or some family member, and 40% are children of single parents. Basically this generation will need to work hard to pay for their undergraduate studies, and if they go to graduate school the debt increases significantly with the added difficulty that in science there is not much job security. After graduation most of the jobs are postdoctoral research positions that are short term because most of them are supported by soft money obtained by the principal investigator that depends on grants. In other words, many postdoctoral researchers need to extend their stay in this category for as long as 6 or 7 years, instead of the usual 3, to obtain an academic position, and that in itself is an elusive task. Women used to be more affected than men, but now the situation is fairly even and a great number of women and men who have trained to the PhD level leave academia, unable to find a promising position.

However, it is important to indicate that in my own view women, based on their biological clocks that determine the importance of motherhood, may be in the unique position to choose between this family path or entering into the competitive market, thus facing job insecurity and the time investment required for an academic researcher to get a permanent position. Saying this, there are women who choose both the academic path and motherhood, but sadly few of them are successful at juggling both as the struggle to raise a family while facing the challenges of finding funding to pursue their research interest is daunting.

The new cadre of cancer researchers needs to find a solution to this problem and help break the barriers that affect the long investments society needs to place in scientific researchers, independently of the gender differences. As we consider a solution to the problem of gender differences in scientific research, we need to accept the premise that during their pre-tenure years women face a limited window of time to start a family, given the knowledge that by the age of 42 only a very small percentage of women are able to conceive naturally; this puts a lot of pressure on the personal life of that young woman scientist already working in an insecure professional environment, competing for short-term contracts and limited grant funding. Therefore, if we want to provide a marketplace free of gender bias we cannot establish a shared path for women and men in science. There should be parallel paths that respect the biology of women and allow them to not only develop their inner biology but also develop their intellectual ambitions at the same time. This will require that our academic environment, using the same metric of quality for both genders, takes into consideration age differences for starting an academic carrier, that it could be later for women than for men. This will require that we as a society accept this as part of our own social network and not as an exception.

It has been suggested that role models could be beneficial for women facing the disjunction of following their biological clock and academic career, and this is fine but it is not a societal solution. There are other suggestions, like additional financial incentives for pre-school childcare support and maternity leave; however, at the moment the majority of the female population is not receiving any support.

Society through its governance has an obligation to provide basic security for maintaining a critical mass of scientific researchers like a long-term investment. This responsibility must include addressing the gender differences. These differences have been a big hurdle in our present society and it is the responsibility of the new cadre of cancer researchers to deal with it. It is clear in my mind that racial differences are not part of this equation because in my view only differences in intellectual capacity should be considered.

6.7. Urban Development and its Impact on Cancer Research

It has been reported that most research activities are concentrated around major metropolitan areas. The top 75 science-producing clusters in the world from 2006 to 2008 generated some 57% of the research, that is, around 3.9 million papers (Van Norden 2010). Among those productive urban areas are Tokyo, London, Beijing, the San Francisco Bay Area, Paris, and New York. Another example of urban influence is Boston; its top research universities with their mammoth budgets have helped to create a vibrant scientific community.

Scientists are attracted to the great urban centers by the freedom to work on their own ideas and the availability of the tools or infrastructure to do so. These basic ingredients have allowed other places in the United States to attract first-grade, innovative scientists. For cancer researchers the presence of a medical center that allows access to patients and also biotechnology incubators where new tools are produced has nurtured the attraction to big urban centers and made a hub for cancer researchers.

However, big urban areas also have significant problems such as transportation, living quarters, and the cost of living, plus pollution issues. Although these challenges can be overlooked, the family of cancer researchers are paying for these conditions and many are not adapting or failing to create the family ties that keep the work as priority over other familial responsibilities.

The new cadre of cancer researchers will need to balance the positive and negative aspects of big urban areas. Globalization, and particularly the digital era, has produced a change in the need for great urban areas as hubs for innovation and scientific research. The fact that the Internet and all the communications media facilitate interaction among scientists makes shared physical space less of a priority than it used to be. Scientists can still perform great scientific discovery in smaller urban places, where social community life and research are compatible. Cities like Philadelphia — which is a great urban area just not of the magnitude of New York, London, or Tokyo — still offer the availability of great medical centers, a critical mass of talented scientists, and places where family and social life are compatible.

6.8. The Social Concern on Environmental Factors and Cancer

There is a great awareness in our society of environmental contaminants and their noxious effects on the body; one main concern is their relation with cancer. These concerns started in 1920 when it was found that tobacco smoking is a cause of lung cancer, although this relationship between tobacco smoking and lung cancer was not widely supported until 1950. In my book *Environment and Breast Cancer*, published in 2011, I presented a multi-author contribution on how environmental factors may also contribute to the development of breast cancer. Dr. Hiatt questions in the first chapter of the book (Russo, 2011), *"What can be concluded about the role of environmental factors, especially about endocrine disruptors and cancer, from epidemiologic studies of humans?"* He indicated that "environment" is frequently defined broadly to include social factors, aspects of the human-built environment, and exposures resulting from so-called lifestyle behaviors (diet and physical activity), but most of the scientific concerns are on the physical environment and exposures such as chemicals, toxins, and radiation. Although tobacco smoking is considered an environmental factor, it is separated from "endocrine disruptors" that are defined by the National Institute of Environmental Health Sciences (NIEHS) as "chemicals that may interfere with the body's endocrine system and produce adverse developmental, reproductive, neurological and immune effects in both humans and wildlife". As it is well summarized by Hiatt, "Endocrine disrupting chemical (EDCs) have effects on male and female reproduction, prostate cancer, neuroendocrinology, thyroid, metabolism and obesity as well as other effects. Some endocrine disruptors that are xenoestrogens (foreign estrogens), may contribute to the development of hormone-dependent cancers such as breast cancer. In many cases positive findings in animal models have stimulated new epidemiologic studies that examine the risks of cancer in humans associated with pharmaceutical and chemical exposures" (Hiatt, 2011). The main concern about the effects of these endocrine disruptors comes not just from animal studies and epidemiological studies, which clearly show a correlation between higher dose exposures in occupational settings and pharmaceuticals that have been associated with biologic/reproductive effects and some cancer outcomes. The best example is the *in utero* exposure to diethylstilbestrol (DES), an estrogen compound given to women between about 1940 and 1970 to prevent miscarriages that resulted in rare cervicovaginal cancers and other reproductive systems abnormalities, including breast cancer (Hiatt, 2011).

My expertise and main contributions in cancer are in breast cancer research. The breast is a fascinating organ from the developmental, physiological, and pathological points of view. As an organ that changes with age and reproductive history of the woman, it makes a target for different pathologies. The breast as an organ absorbs and secretes what is in the environment. Meaning any substance that reaches the circulating blood of a woman is filtered in the breast and is either secreted in the milk of lactating women or accumulated in the stroma and epithelia of the glandular structure of this organ. It is calculated that one out of eight women will develop breast cancer and that approximately 45,000 women die yearly from this disease. The economic impact, in terms of value of life lost from breast cancer, is in the hundreds of billions of dollars and because it strikes women while they are in their most productive years and has myriad influences on the quality of life and impact on families, it has a huge impact on women's overall health. Therefore, it is reasonable to correlate the incidence of this disease in industrialized nations with environmental factors. In 1982, I published a seminal work indicating that in the life span of a woman there are different periods of susceptibility, pointing particularly to pregnancy occurring early in life as preventing breast cancer whereas the other period that can be associated with increased risk is if the human breast is exposed to a carcinogenic insult (Russo *et al.*, 1982). This has originated the concept of windows of susceptibility, such as *in utero,* in childhood before puberty, during adolescence and breast development, the period leading up to a first pregnancy, and during pregnancy itself. In all these phases of the female life-course, the breast, which is the only organ to develop after birth, is undergoing some stage of development during which environmental agents may have their effects. Much of the evidence for a relationship between the chemical environment and cancer comes from exposures in occupational or therapeutic settings. As Hiatt indicated, "lessons have been learned from exposure to pharmacologic doses of exogenous estrogens (i.e., diethylstilbestrol [DES]) exposure in utero or from the accidental exposures of workers to dioxin".

It is well-known that a lower age of first full-term pregnancy confers substantial protection against postmenopausal breast cancer compared to first full-term pregnancy at a later age (the cutoff is usually around 35 years), or nulliparous women (Hunter *et al.*, 1997)]. This is thought to be because during pregnancy the breast cells and breast architecture are fully mature, and there is less proliferative activity, and thus decreased sensitivity to environmental carcinogens (Russo *et al.*, 2001). Parity is also directly associated with a protective effect as each successive live birth confers additional protection, presumably by the same mechanism. However, pregnancy also transiently increases the risk of breast cancer in the first 5 years or so after delivery (Williams *et al.*, 1991). Thus the event of pregnancy and its frequency represent another window of susceptibility during which the human breast is at risk of environmental carcinogenesis, potentially including risks associated with exposure to environmental agents. A systematic review of environmental pollutants and breast cancer has been published by Brody *et al.* (2007). Among the agents that have been related to cancer is DES exposure. This estrogenic compound was taken off the market when daughters of women who took it during pregnancy developed adenocarcinomas of the vagina and subsequently, as the women who took DES have aged, multiple reproductive abnormalities have emerged, including increased breast cancer rates.

It has been described by Hiatt:

> Other environmental agents are derived from the organochlorines, that is, a group of synthetic chemicals, some of which have properties of endocrine disruptors. They include polychlorinated biphenyls (PCBs), dioxins, and organochlorine pesticides such as DDT, lindane, heptachlor, dieldrin, aldrin, and hexachlorobenzene and triazine herbicides such as atrazine. These pollutants are lipophilic and resistant to degradation and are persistent in individual fat stores, the food chain, and the environment and have become highly prevalent in industrialized countries since World War II. However, the association of organochlorines with breast cancer has been inconsistent. Other prevalent environmental agents are polycyclic aromatic hydrocarbons (PAHs), a large class of chemicals formed by the incomplete combustion of coal, oil, and gas, as well as grilled meats, tobacco smoke, and other substances to which humans are exposed in ambient air. These substances are genotoxic and known to be potential breast carcinogens. Another widely used industrial monomer is bisphenol A (BPA), a weakly estrogenic chemical that is polymerized in the manufacture of polycarbonate plastic and epoxy

resins. Parabens are antimicrobial preservatives found in personal care products, including underarm cosmetics, that can act like weak estrogens to bind to the estrogen receptor. Personal care products and cosmetics may include endocrine disruptors in the categories of phthalates and organic solvents as well as parabens. A study of brominated flame retardants (polybrominated biphenyls, PBBs) among accidentally exposed farm workers in Michigan revealed an association with earlier pubic hair, but not breast development, in the daughters of exposed mothers. Polybrominated diphenyl ethers (PBDEs) are a subgroup of flame retardants, which are now being phased out because of proven toxicity, although a number are still in common use and under study in relationship to pubertal development. Although no epidemiologic studies have been done in adults, in a study of six congeners of PBDEs serum levels were detected in 70% of the sample of girls from California and Ohio. Levels were higher for girls from California and for black girls compared to white girls. Perfluoroalkyl acids (PFAAs) are a family of perfluorinated chemicals that have also found their way into the environment and have endocrine-disruptor properties. Two of the most widely studied are perfluoroactanoic acid (PFOA) and perfluorooctane sulfate (PFOS). Another prevalent environmental agent is phytoestrogen. Phytoestrogens in food have weak estrogenic or antiestrogenic effects and fall into three major classes: isoflavones, lignans, and coumestans. Isoflavones are found mainly in soy products and are largely in two forms, genisten and diadzein, with glycetin a third, less common variety. Lignans are found primarily in grain products. Soy intake is substantially higher among Asian populations and since dose is a critical factor, discussion of epidemiologic results for isoflavones needs to take the population studied into account. (Hiatt, 2011)

All of these data clearly indicate that environmental agents are in our daily life, and although some of them have been shown to be carcinogenic in experimental animals or promoting endocrine abnormalities leading to cancer, the extrapolation of those data to humans has not clarified the actual risk to humans from these compounds. Major problems include the difference in exposure between animals and humans, which creates a great uncertainty in the role of the environment and cancer. The new cadre of cancer researchers needs to face this uneasiness created by the lack of definitive data and end the quarrel between industry and lobbyist.

6.9. Is There a Utopian Society for Performing Scientific Research?

In 1516 Sir Thomas More coined the term "utopian society" to describe a fictional island society in the Atlantic Ocean. In his book More refers to a community or society that, in theory, possesses highly desirable or nearly perfect qualities. Among these qualities are egalitarian principles of equality in economics, government, and justice. Since the publication of this book the term has been used to describe both intentional communities that attempt to create ideal societies and imagined societies portrayed in fiction. When this concept is applied to scientific research, it is expected that in the future advanced science and technology will allow utopian living standards, e.g., the absence of death and suffering; changes in human nature and the human condition. In a certain way, humanity has advanced several quantum leaps since its emergence on this planet, and from the biological and medical points of view, we should expect changes that can improve our living conditions. The best proof of this is that science has affected the way humans live to such an extent that in every period our species has been pushing the edge for a better life. In the conquest of cancer, scientific research has been incremental but the pace is getting faster and faster. In the following paragraphs, I make a brief summary of how scientific research has brought us to the present stage of knowledge and how we can go forward.

6.9.1. *Greco-Roman period*

Scientific inquiry in classical antiquity was also aimed at curing diseases that were afflicting the different strata of society. Although these scientists called themselves natural philosophers, they worked as physicians and healers. But Plato and Aristotle were the ones who produced a systematic discussion of the natural events. One of their most important contributions was the empiricism indicating that universal truths can be arrived at via observation and induction, thereby laying the foundations of the scientific method. Aristotle made significant advances in biology and produced many writings on biological causation and the diversity of life. He made countless observations of nature, especially the habits and attributes of plants and animals in the world around him, classified more than 540 animal species, and dissected at least 50. Aristotle's findings influenced subsequent Islamic and European science (Lyons and Petrucelli, 1978; Carmichael and Ratzan, 1991).

The important legacy of the Greco-Roman period includes substantial advances in factual knowledge, especially in anatomy, zoology, botany, and mathematics. In a period of 600 years, three important scientists emerged and they established the medical system of today. Hippocrates (circa 460–377 BC) was the first to describe many diseases and medical conditions and also the first one to recognize cancer as a different disease. He developed the basic principles of how to practice medicine. The second was Herophilos (335–280 BC), the first to make conclusions for diseases by dissection of the human body and he has even been attributed with describing the rudiments of the nervous system. Lastly, Galen (circa 129–200 AD) established not only the concept of clinical medicine but ventured into the use of surgery, including brain and eyes, which was not attempted again for many centuries afterward.

6.9.2. Renaissance

The renaissance in medicine and biology took place from 1400 to 1700 of the current era and it was based on Greek and Roman learning. However, an important development that helps explain the explosion of knowledge was the invention of movable type, which made printing books much easier and cheaper. This period produced two pivotal contributions: the study of anatomy and the invention of the microscope. The printing press made possible the diffusion of medical ideas and anatomical diagrams and also a diffusion of the knowledge of Galen. Among anatomists who have made a significant contribution to our medical knowledge are Paré, Vesalius, and Harvey (Lyons and Petrucelli, 1978; Carmichael and Ratzan, 1991).

Paré was a French surgeon and greatly curious about the human anatomy. He was a military surgeon during the French campaigns of 1533–1536 in Italy and invented many surgical instruments. Paré even turned to an ancient Roman remedy using turpentine, egg yolk, and oil of roses to relieve the pain of wounds and found that it sealed them effectively. The second scientist that made important advances in our knowledge of anatomy was Andreas Vesalius (1514–1564). He was a Flemish-born anatomist whose dissections of the human body helped to rectify misconceptions dating from ancient times, particularly those of Galen, who only studied animals such as dogs and monkeys. He wrote many books on anatomy from his observations, most notably *De Humani Corporis Fabrica*, which contained detailed drawings of the human body posed as if alive. William Harvey (1578–1657) was an English medical doctor-physicist, known for his contributions in heart and blood movement.

As part of the Renaissance, we must acknowledge the development of the microscope. Evidence points to the first compound microscope appearing in the Netherlands by the 1620s. Galileo in 1625 developed the compound microscope called the *occhiolino* or "little eye". But it was Antoni van Leeuwenhoek in 1676 that really applied the use of microscope for observing microorganisms and with that the field of microbiology was born.

6.9.3. *The Industrial Revolution and its great medical discoveries*

It was not until the 19th century that the practice of medicine changed, revolutionized by advances in chemistry and laboratory techniques and equipment; old ideas of infectious disease epidemiology were replaced with bacteriology and virology. In the mid-1800s, Pasteur started making a systematic study to demonstrate that microorganisms were not only living beings responsible for the fermentation produced by yeast but also that they could be the carrier of diseases. Koch published a landmark treatise in 1878 on the bacterial pathology of wounds and in 1881; he reported the discovery of the "tubercle bacillus", cementing germ theory.

Paris and Vienna were the two leading medical centers on the European continent in the years 1750–1914. From the 1770s to the 1850s, Paris became a world center for medical research and teaching. The "Paris School" emphasized that teaching and research should be based in large hospitals and promoted the professionalization of the medical profession and an emphasis on sanitation and public health. Louis Pasteur (1822–1895) was one of the most important representatives of French science at that time. The First Viennese School of Medicine, 1750–1800, was led by the Dutchman Gerard van Swieten (1700–1772), who aimed to put medicine on new scientific foundations, promoting unprejudiced clinical observation, botanical and chemical research, and introducing simple but powerful remedies. After 1871, Berlin, the capital of the new German empire, became a leading center for medical research and its most outstanding researcher was Robert Koch.

The American Civil War (1861–1865) had a dramatic long-term impact on medicine in the United States, from surgical techniques to hospitals to nursing and research facilities. By the late 19th and early 20th centuries, English statisticians, led by Francis Galton, Karl Pearson, and Ronald Fisher, developed mathematical tools such as correlations and hypothesis tests that made possible much more sophisticated analysis of statistical data. European ideas of modern medicine were spread widely through the world by medical missionaries and the dissemination of textbooks. Japanese elites enthusiastically embraced Western medicine after the Meiji Restoration of the 1860s. The Great War spurred the usage of Roentgen's X-ray and the electrocardiograph for the monitoring of internal bodily functions. This was followed in the interwar period by the development of the first antibacterial agents such as the sulfa antibiotics.

Global health and life expectancy in most of the world have improved since 1948, to well above 80 years in some countries. Eradication of infectious diseases is an international effort, and several new vaccines have been developed during the postwar years, against infections such as measles, mumps, several strains of influenza, and human papilloma virus, though the early success of antiviral vaccines, antiviral drugs, and antibacterial drugs was not achieved until the 1970s. Through the WHO (World Health Organization), the international community has developed a response protocol against epidemics. As infectious diseases have become less lethal, the most common causes of death in developed countries are now cancer and cardiovascular diseases; accordingly, these conditions have received increased attention in medical research. Cancer treatment has been developed with radiotherapy, chemotherapy, and surgical oncology.

6.9.4. *The 21st century*

In the 21st century, we are still struggling with cancer and also infectious diseases, as well as degenerative ones like Alzheimer's and cardiovascular diseases. In this section, however, I have concentrated only on outlining the major advances in cancer research that provide great potential for the prevention and treatment of cancer. Most of these topics have been analyzed in detail in Chap. 2 (Trends in Scientific Discovery), in which I discussed the role of chromatin remodeling, variations in the genome, new advances in immunology, and mainly the checkpoint blockade, synthetic biology, and the new models in cancer research.

6.9.5. *Final considerations*

As our brief historic review indicates, there is not a utopian place for performing cancer research and we as humans struggle continuously with old diseases as well as new complications of the diseases that we create because of our ignorance. We have advanced enormously when compared with our humble beginnings, but social and economic conditions have created a difficult milieu for moving forward. Probably we need to continue facing the present and the future of cancer research, integrating new factors into the equation with the hope that we can alleviate cancer suffering, like we as scientist have done in the past for other illnesses.

References

Alberts, B. On effective leadership. *Science*, **340**: 660, 2013.

Brody, J.G., Moysich, K.B., Humblet, O., Attfield K.S., Beehler, G.P., and Rudel, R.A. Environmental pollutants and breast cancer: Epidemiologic studies. *Cancer*, **109**: 2667–2711, 2007.

Carmichael, A.G. and Ratzan, R.M. Medicine. A treasury of art and literature. New York: Harkavy Publishing, 1991.

Gachter, S. and Schultz, J.F. Intrinsic honesty and the prevalence of rule violations across society. *Nature*, **531**: 496–499, 2016.

Hiatt, R.A. Epidemiologic basis of the role of environmental endocrine disruptors in breast cancer. In *Environment and Breast Cancer*, J. Russo (ed.), pp 1–28. New York: Springer, 2011.

Huang, A.S. Passions. *Science*, **344**: 1362–1365, 2011.

Hunter, D.J., Spiegelman, D., Adami, H.O., van den Brandt, P.A., Folsom, A.R., Goldbohm, R.A., Graham, S., Howe, G.R., Kushi, L.H., Marshall, J.R., Miller, A.B., Speizer, F.E., Willett, W., Wolk, A., and Yaun, S.S. Non-dietary factors as risk factors for breast cancer, and as effect modifiers of the association of fat intake and risk of breast cancer. *Cancer Causes Control*, **8**: 49–56, 1997.

Hwa Y.J. Comparative Political Culture in the Age of Globalization. Rowman and Littlefield, 2008.

Lyons, A.S. and Petrucelli, R.J. Medicine: An Illustrated History. New York: Harry N. Abrams Inc., Publishers, 1978.

Nurse, P., Treisman, R., and Smith, J. Building better institutions. *Science*, **10**: 341, 2013.

Russo, G. For love and money. *Nature*, **465**: 1104–1107, 2010.

Russo, J. Environment and breast cancer. New York: Springer, 2011.

Russo, J., Hu, Y-F., Silva, I.D.C.G. and Russo, I.H. Cancer risk related to mammary gland structure and development. *Microsc. Res. Tech.* **52**: 204–23, 2001.

Russo, J., Tay, L.K. and Russo, I.H. Differentiation of the mammary gland and susceptibility to carcinogenesis. Breast Cancer Res. Treat. **2**: 5–73, 1982.

Suresh, S. Cultivating global science. *Science*, **336**: 959, 2012.

U.S. Equal Opportunity Employment Commission, www1.eeoc.gov/eeoc/statistics/employment/jobpat-eeo1/2009/index.cfm#select_label.

Van Norde, R. Building the best cities for science. *Nature*, **467**: 906–908, 2010.

Williams, E.M., Jones, L., Vessey, M.P., and McPherson, K. Short term increase in risk of breast cancer associated with full term pregnancy. *BMJ*. **300**: 578–579, 1990.

Wolff, M.S., Collman, G.W., Barrett, J.C., and Huff, J. Breast cancer and environmental risk factors: Epidemiological and experimental findings. *Annu. Rev. Pharmacol. Toxicol.* **36**: 573–596, 1996.

Yang, W. 2011. Paper Tiger. *New York Magazine*, May 8. http://nymag.com/news/features/asian-americans-2011-5/.

Further Reading

Andrew, F. Global collaboration. *Nature*, **482**: 122, 2012.

DeWeerdt, S. The urban downshift. *Nature*, **531**: S52–53, 2016.

Gilbert, N. A natural high. *Nature*, **531**: S56–57, 2016.

Maxmen, A. The privilege of health. *Nature*, **531**: S58–59, 2016.

Pukkala, E., Kesminiene, A., Poliakov, S., Ryzhov, A., Drozdovitch, V., Kovgan, L., Kyyrönen, P Malakhova, I.V., Gulak, L., and Cardis, E. Breast cancer in Belarus and Ukraine after the Chernobyl accident. *Int. J. Cancer.* **119**: 651–658, 2006.

Russo, I.H., Russo, J. Mammary gland neoplasia in long-term rodent studies. *Environ Health Perspect.* **104**: 938–967, 1996.

Wolff, M.S., Britton, J.A., Boguski, L., Hochman, S., Maloney, N., Serra, N., Liu, Z., Berkowitz, G., Larson, S., and Forman, J. Environmental exposures and puberty in inner-city girls. *Environ. Res.* **107**: 393–400, 2008.

Chapter 7

The Perception of Cancer Research by the Public

7.1. Introduction

Cancer research is one of the few areas of science in which the public, as activists, lobbyists, volunteers, and patients, have transmitted their opinions and views as well as their wisdom to the scientific and medical community. In the case of breast cancer, a major shift took place in 1980 when former First Ladies Betty Ford and Nancy Reagan, along with the founder of the Susan G. Komen Foundation, Nancy Brinker, began speaking publicly about the personal impact of the disease, which increased awareness of breast cancer and made it more acceptable to talk openly about the disease. The open discussion of the impact of breast cancer on the lives of these distinguished women was also associated with a better understanding of the statistics about breast cancer (Braun, 2003). Other fighters, like Rose Kushner, the founder of the National Breast Cancer Coalition, and Frances Visco, have created a movement of advocates that allows consumers to engage in the process. Not only do the advocates aim to educate women about the importance of breast self-examinations and screening mammograms and clinical breast examinations, but also they lobby the government for funds to fight this disease, like in the case of the Department of Defense allocation of funds for breast cancer research and now for other types of cancer, as well as the creation of private funding like that implemented by the Komen and Avon Foundations. The advocacy movement in breast cancer provided a road map for engaging the business, government, and scientific communities as partners in advocacy. It was followed for other cancer types and the lessons learned have created increased awareness in public and private sectors as well as spurred the government to take action on their demands.

7.2. An Example that Deserves to be Imitated

An article written by Eric Hand and published in *Nature* describes an interesting story of how the public aided researchers by volunteering to help in large scientific projects. A computing project called Rosetta@home allowed volunteers to download a small piece of software and let their home computers do some extracurricular work when the machines would otherwise be idle. The idea behind this project was to study protein folding. Thousands of people signed up for Rosetta@home and provided their findings to the researcher, who emerged with entirely new folding strategies. It seems that this approach has the most promise in areas such as vision, language, and complex logic puzzles — territories in which humans are expected to retain an edge on computers (Hand, 2012). Another interesting example in which volunteers from the public were enlisted in a research project was a cosmic-ray experiment called TREK. Basically this project consisted of especially "designed glass plates mounted on the outside of the Russian space station Mir in 1991. Cosmic-ray particles pelting the glass left microscopic traces that were revealed by chemical etching after the TREK detector had returned to Earth in 1995". The researcher used the automatic imaging microscope from TREK to create 1.6 million images of the aerogel. The principal investigator in this project estimated that it would take a century for one person to peruse them all. For this purpose he created Stardust@home, a continuing project that enlists the pattern-recognition abilities of thousands of volunteer "dusters". The results of these studies were later published in *Nature*.

Programs similar to Stardust@home, such as Hominids@home, have been created and many researchers are using this approach involving volunteers that facilitate discovery and provide important leads in projects that require a significant amount of workforce. A recent example is the discovery of the *Homo naledi* in South Africa by Lee R. Berger, who used hundreds of volunteers in the lead-up to this important discovery.

Although not all members of the scientific community embrace the strategy of using volunteers for research projects, the evidence indicates that if the principal investigator leading the project canalizes the energy of laypeople, their common logic and wisdom benefit the project significantly. In 2012, as a Director of the Breast Cancer Research Laboratory at Fox Chase Cancer Center in Philadelphia, Dr. Irma H Russo and I established an agreement with Drexel University to have, through their co-op program, undergraduate students that rotate for 6 months *ad honorem*, working 20 hours a week in breast cancer research. In this internship, the students are engaged in the collection of data from *in vitro* and *in vivo* studies; perform procedures such as histological sectioning of paraffin-embedded tissue; perform immunohistochemistry work; perform digital recording of images; carry out quantitation of data using different software applications, data entry, and statistical analysis; use bioinformatics engines and programs; perform analysis of data and interpretation; do scientific literature searches; present results at lab meetings and assist in writing research articles; and perform other duties related to general lab operation to be shared with team members. These students provide a significant intellectual force that has been easily integrated into our research endeavor and more than 60 undergraduate students have been trained in cellular and molecular biology as well as in bioinformatics. This is an example of how the utilization of human resources accomplishes two important objectives: training and guiding young people in cancer research and at the same time obtaining invaluable help in the research process, which otherwise would require significantly more economic resources and time to complete.

7.3. The Public Must be Aware of the Role of Women in Cancer Research

Although in the United States and Europe, around half of those who gain doctoral degrees in science and engineering are female, only one-fifth of full professors are women. This disparity is shocking, but the solution is at hand. Simply creating adequate childcare programs would help women fulfill their dual roles. There is also another related problem that should be considered as there is a relatively simple solution. That is, the offering of job opportunities not only to the male or female candidate but also to his or her partner; this issue has been addressed in the literature (Holmes, 2012).

7.4. The Use of Patenting and its Effect on the Public

The recent approval of The America Invents Act (Reich, 2011) has been considered the culmination of a long campaign by universities and companies to change the US patent system to a first-to-file arrangement, in line with most of the rest of the world. Independent of the legalities that many are discussing concerning the wisdom of this act, it will provide adequate protection to the inventor and researchers that invested in developing an idea, a product, or a device.

One important factor that needs to be considered in the equation is the public. How is the public affected by the patenting of a product and the restrictions placed on its wider use? While patenting is undoubtedly the seal of creativity, there is concern about applying this concept to biology, mainly to inventions relating to medicine. A clear demonstration of these feelings is the US court decision ruling that patents on two genes linked to ovarian and breast cancer, *BRCA1* and *BRCA2*, were illegal; the court went even further, indicating that the claims of ownership were in violation of a "law of nature". The practical part of this ruling is that the test can be done by other parties at a reduced cost for the patient. This precedent also impacts the patenting of 2000 other genes in the pipeline.

7.5. How the Public Can be Integrated into the Scientific Arena

Advancement in cancer treatment would not be possible without the participation of volunteers in clinical trials and this is not possible without informed consent. The notion of informed consent, that is obtaining permission from educated volunteers, can be traced back to research-ethics principles introduced in response to revelations about the way Nazi doctors tortured people during World War II. Research as well as funding institutions are implementing the regulations to protect the use of human subjects in cancer research. Decades ago, researchers did not imagine how much information could be obtained from a piece of stored tissue. Even now, it is difficult to conceive of the possibilities for obtaining an extraordinary amount of data, and even more difficult to explain to a research subject all the possible ways in which their offered data might be used in the future. The underlying problem is that it is becoming less possible to guarantee privacy, even under the strict rules that mandate the deposition of research data in public databanks. The public is becoming more aware that confidentiality can be broken and their personal data are vulnerable. This is creating a dilemma not only for the researchers but also for the volunteers who participate in a clinical trial or large data analysis. Although the issue is under consideration by scholars of ethics and law, who are searching for new models of informed consent that could accommodate the needs of researchers and research participants, the outcome is not at hand. However, a sobering thought is that the public must be involved in the cancer research process and participants must be aware of the difficulties as well as the bad and good outcomes of the studies performed on human subjects. The emphasis on the benefits of scientific research for public well-being must demonstrate that active participation in solving medical problems is not a one-way street but a civic duty. When both researchers and volunteers understand these basic principles, many of the problems that we are creating will simply go away. But if controlled access is not the right solution, it is up to the research community, in consultation with the public, to devise a better one. A solution should come sooner, rather than later, because this latest revelation of a privacy loophole will certainly not be the last.

An important principle in this rapidly evolving technology is that the public does not remain behind the main advances in science. It would be a mistake to allow the societal implications to be managed by experts only. Sarewitz (2010) has eloquently addressed this issue indicating that "We are an innovating species, engaged in a balancing act. In the decades after the Second World War, innovation fuelled an unprecedented era of wealth creation while keeping us on the brink of nuclear annihilation". The basic problem is that once a complex technology is widely used, restricting, reorienting, or replacing it becomes incredibly difficult. Therefore, the interaction between the public and the innovators must be opened to discussion early on in the process. It would be a mistake for cancer researchers to ignore the public perception of their science because the reality is that neither the public nor the scientist can predict the outcome of a clinical trial, but it is a healthy interchange among the parts with great potential benefits. An example in which this cross talk between scientists and the public as well as the government is necessary is in the use of stem cells. Although the government has posed regulation not only in the United States but also in Europe, a healthy discussion of the pros and cons of the use of fetal tissue for stem cell research must continue, because what science sees as beneficial today could be outdated thinking tomorrow. The history of medicine is full of mistakes like the use of radical surgical methods for treating cancer in the 1960s and 1970s, which eventually demonstrated that radical surgery was not the best way to proceed. Although these mistakes are unavoidable in the evolution of our knowledge in the treatment of cancer, it is also a good precedent to listen not only to history but also to ponder other alternatives that reflect the public opinion. For that, of course, the laypeople need to be involved. The use of an "ethics panel to launch a discussion about a new technology sent a good signal: the government's role is not just to throw money at the next big thing, but to encourage open talks about social implications and options" (Sarewitz, 2010). This is a good approach, but we cannot stop there and social media may help facilitate a broader, healthier discussion.

7.6. The Scientist, the Public, and the Government

The scientists, the public, and the government are all integral parts of our society and when scientific decisions affecting the public are made by politicians, it is the responsibility of the scientists to be sure that they have articulated their position clearly. This message is targeted to cancer researchers who are aiming, e.g., to gain government approval of a specific new health measure. In my experience, it is important to take a position of mutual respect because whatever is the high view of our work, as scientists we sometimes fail to appreciate the complex, multifarious nature of decision-making and that the end point is the public. Although it is not an immediate solution to this problem, it is important that scientists consider the well-being of the public because after all we are working to save lives and benefit people. Therefore, it is a responsibility of us, as scientists, to understand politicians and interact with them by improving communication and encouraging our youngest scientists to do the same. Finally, we must all take part in the political process, presenting our diverse views.

These three components of our society, the scientists, the public, and the government, are also part of an entire issue on universal access to needed health services without financial hardship in paying for them. But the reality is that universal health coverage is faced with almost limitless need and finite resources due to aging, growing populations, and ever-more-sophisticated and expensive technologies. In an article in *Nature*, Shelton (2013) elaborated on these concepts, indicating "that although clinical services clearly have benefits, but their cost-effectiveness and impact on population-level health are far from clear. That is partly because therapies are not always effective and some even harm. But most important is that medicine's curative arsenal tends to arrive too late to address the drivers of disease". Although it is extremely difficult to find a solution for developing economies, in the developed ones the emphasis still needs to be shifted from payment for procedures to the prevention of diseases. I will address breast cancer as an example. Breast cancer prevention in the 21st century must be adjusted to a complex strategy with an emphasis on childhood and window of opportunity. In addition, focus on etiology of different molecular types and identification of high-risk women would improve our diagnostic and preventive capabilities. Since susceptibility originates during childhood, increased attention must be given to risk factors that induce genomic imprints and cause cancer later in life. Such conditions are obesity, the related early sexual development (puberty, menarche), lifestyle, environmental toxins, radiation, physical activity, etc., as has been discussed by Jaak Janssens. The author stresses that a "healthy lifestyle should be implemented as early as the postnatal period but has benefits if started up to at least to the third decade of life. In addition to instituting some protective measures during childhood, interventions during childhood and early adulthood could also be useful. The persisting thought that adults can only slightly reduce risk might not be completely true. Certainly postmenopausal breast cancers could be reduced by proper treatment of the metabolic syndrome through common medications such as statins and oral antidiabetics. Alcohol consumption should be moderate to low, at least for some genetic pleomorphism. Chronic anti-inflammatory medication could provide some protection as well. Screening, with attention to lower social classes and developing countries, along with awareness campaigns should give an answer to the steep rise in breast cancer. Digital mammography seems still to be the first choice and affordable to larger populations. Further research is mandatory but difficult to steer". As Janssens has indicated, "overlooking the priorities one should aim at prevention during childhood and adolescence, the use of molecular intelligence, but most importantly to look at screening with affordable means. Sophistication of medical technology is no longer the ultimate and should be replaced by concentrating on easily implementable and affordable technology that can be used to address the breast cancer plague in developing countries" (Janssens, 2016). Support for this contention is excellently described in the book *Next Medicine: The Science and Civics of Health* by Bortz II (2011). In this book, the author argues that it is time to turn to a "next medicine". Its first principle would be the recognition that medicine's mission is "to assert and assure human potential". The new medicine would entail practice that focuses on the whole person (not her or his parts). He proposes moving a substantial portion of current health-care investment toward health education, behavior modification, and disease prevention. He claims that doing so will be cost-effective, yielding both great reductions in expenditures and improvements in health. In describing "health", Bortz draws from his research and practice in aging and the roles physical fitness has played in his own and others' lives. He believes that "we have the potential for far greater spans of life and quality of living than most people realize". Perhaps the book's most controversial proposal is the recommendation to incentivize personal health behavior through risk-adjusted premiums. Therefore, the new cadre of cancer researchers must create an adequate balance for furthering medical solutions for the ailing world.

7.7. Public Perception

There are striking differences in views on science-related issues between the public or society and the scientists. For example, one issue that is of concern to the public is the safety of eating genetically modified foods; most scientists are not concerned on the issue. The use of animals in research is also not well understood by the public, whereas it is of no concern for the scientists. On the other hand, scientists are concerned that the public eats products that have been treated with pesticides; conversely, more than 70% of the population is not worried. The same applies to the concern of scientists that weather changes are caused by human activity, yet not all the public believe that those concerns are justified. These differing opinions create a disparity and a significant gap between how scientists think and how the public perceive it. Meaning that the scientists are not delivering a clear message that needs to be repeated over and over again until the public or society accept the facts that science is providing them.

The scientists' messages lack impact on the citizenry in part because the confidence that the public has in scientists has been eroded, in part due to fraudulent papers that have appeared in relevant scientific journals. For example, prominent journals like *Nature* needed to retract the paper on the isolation of human stem cells, and when the media took over the disaster was evident. Although these are real cases, most of the damage is done by the media, which at times portrays scientists as the "bad guys". It is easy to create a stereotype of heroes crusading for the truth, and that's fine, but it doesn't make scientists the villains. Yet there is no easy way to correct this lingering and unjust stereotype, but a continuous presence of scientists in public media will eventually demonstrate the positive role that they have on our well-being.

7.8. Educating the Public

Radio and television have played an important role in educating the general public in matters related to science and medicine. What the public does not know firsthand is that each of these educational programs requires months of preparation.

The educational material to be displayed either in the radio or in television must be crispy and clear enough to be understood. The corollary of this educational material is that to be productive, it must be continuous; it cannot be a sporadic event but must repeat weekly or monthly to permeate the public.

The final question is why these educational programs must be done at all, and the reality is that the use of the media to educate the public is probably one of the best ways to advance our society and also make the science better understood by the common citizen. Therefore, all the energies that these producers invest in these programs are a great service whose value can only be measured in time. The public must be aware and provide feedback to those producers that are taking this task. The new cadre of cancer researchers must cooperate with and promulgate these events. The more informed and better educated the public is on science and medical matters, the better cooperation will be with the cancer research community. The effort to bring educational material to the public will definitely make the scientific findings transparent and can result in a positive strength to foster cancer research.

These educational programs must show how the scientific data are obtained and analyzed; this spirit of openness is vital for showing the difficulty in probing the value of a drug, for example, or illustrating the need for obtaining consistent and reproducible data. This openness in how scientists work will create a consciousness of how new discoveries come about. It is important that in this process the cancer researchers be extremely careful not to introduce data or scientific facts that make the science vulnerable or demagogic; some believe this approach is the way to build a reputation in the community. It is on this point that the new cadre of cancer researchers needs to be extremely careful; they must determine which information is ready to be discussed without creating false hope or misleading the public.

For example, psychology has been rocked by irreproducible findings in the past few years, and they have been especially vocal about broadcasting findings that are not reproduced in other places. This does not indicate that the scientists generating these data are wrongdoers but rather that in dealing with human behavior more strict rules of reproducibility must be established before findings are made known to the public. This also applies to the use of drugs that are not properly studied and revealing data collected without following the standard protocols. These kinds of data are confusing and could be dangerous if they are not used with caution.

Whereas radio and television are the most frequent sources of public education, now the Internet is taking on a major role but still it is not the main source for general broadcasting. Another method that I advocate is the developing of conferences targeted to the public. These conferences are already helping many communities to be enriched. There are conferences on general science or medical points of interest organized by hospitals or free libraries that are helping the public. One of the shortcomings of this method of communication is the low visibility and availability that limit attendance by the general public. Another factor is the length of these lectures — as a rule about 18–20 minutes are enough to transmit the message. Lectures that are open to the public but extend over a lengthy period of time do not entice attendance or the attention of the listeners.

The speaker must be able to translate his or her scientific knowledge in a creative manner without the need for many graphical aids or notes. It is important that the speakers provide a freely flowing, relaxed presentation style, without notes. It is also not a bad idea to request a small admission fee by the organizers; this would allow an adequate presentation environment and a way to pay for acquiring new knowledge. Scientists wishing to inspire nonscientists can use these approaches to entice an audience. As discussed in Chap. 6, the ability to explain science like a storyteller is probably the best way to educate and make the public interested in science.

7.9. The Public is Afraid of Losing their Privacy

Is this statement really true? The answer is yes. Evidence is piling up that our privacy is getting less private for every day that passes. I am not talking about birth certificates, health insurance, credit cards, bank accounts, or even social security numbers, but things like a smart phone always knows where you are. If you add to the smart phone, other social media outlets such as Facebook, LinkedIn, Twitter, and e-mail, you create a public identity for yourself. If your genome data are collected, the chances are great that those data will be shared and eventually compared with your clinical history, making it almost certain that you can be tracked down. Although some of these are sophisticated methods of identification, other technology is currently in progress, like face recognition and new computational techniques that can identify people or trace their behavior by combining just a few bytes of data that make them an extremely visible, non-private identity. The reality of losing privacy could be considered inconsequential in a utopian society, but unfortunately we are not living in a utopian society.

What to do with this conundrum? There is no solution at hand. Medical researchers try, through informed consent and HIPAA (Health

Insurance Portability and Accountability Act) regulations, to protect the identity of their patients but even this cannot safeguard the individual data or eliminate ways to trace the individual behind this mega database. The overarching solution is to build a better society in which individual data are considered private even though they could be identified. However, efforts to protect privacy by controlling use of data are gaining more attention and this has been discussed in an article recently published by Landau (2015) in *Science*.

References

Bortz, W.M., II. *Next Medicine: The Science and Civics of Health*. Oxford: Oxford University Press, 2011, p. 265.

Braun, S. The history of breast cancer advocacy. *Breast J.* (Suppl 2), S101–103, 2003.

Hand, E. People power. *Nature*, **466**: 685–687, 2012.

Holmes, M.A. Working together. *Nature*, **489**: 327–328, 2012.

Janssens, J. The paradigms in breast cancer prevention. In: *Trends in Breast Cancer Prevention*, Jose Russo (ed.), pp. 1–21, Chapter 1, New York: Springer, 2016.

Kassen, R.K. If you want to win the game, you must join in. *Nature*, **480**: 153, 2011.

Landau, S. Control use of data to protect privacy. *Science*, **347**: 504–506, 2015.

Reich, E.S. US legislation aims to simplify rules for inventors. *Nature*, **472**:149, 2011.

Sarewitz, D. Not by experts alone. *Nature*, **466**: 688, 2010.

Shelton, J.D. Ensuring health in universal health coverage. *Nature*, **492**: 453, 2013.

Chapter 8

Measuring Scientific Innovation in Cancer Research

8.1. Introduction

In the decades before and after 1900, health care was revolutionized by advances in disease prevention, surgery, and drug treatments that allowed management of chronic afflictions. One notable medical innovation was aspirin. But nothing captured the headlines like the extraction of insulin in 1921 by Dr. Frederick Banting and Charles Best, the development by Louis Pasteur of a rabies vaccine in the 1880s, and the 1895 discovery of X-rays by Wilhelm Röntgen. Remarkable advances in our knowledge of the chemistry of life and biotechnology achieved in the past century have created a tremendous task: We must understand biological networks in order to clarify, and in some cases redesign, our concept of cancer. The challenge is understanding how all the thousands of proteins in a living cell interact with each other. Of the approximately 21,000 distinct proteins encoded by the human genome, we have a good knowledge of the function of less than half of them, and far less understanding of how they work in normal individual cells — less still do we know about how they work in the cancer cells. In front of this challenge is the opportunity for biologists, biochemists, mathematicians, computer scientists, and engineers to develop innovative research strategies that make sense of the enormously complicated network of molecular interactions found in even the least complex living cells. These advances are necessary if we want to understand and conquer cancer.

The problem is how to do it. The first obstacle that the cancer researcher has to surmount is how to determine innovation when still facing unsolved basic biological problems. How can we measure the impact and significance of our research and its translational value in a germinal idea? How can the biomedical science establish biomedical relevance when all the problems that we face in cancer research are relevant? True innovation emerges from understanding the biological processes and that is what we should aim for in cancer research. If we do not, innovation, impact, significance, and translational medical relevance will lose their meaning because they are all the consequences, not the driving forces, of the scientific inquiry.

8.2. Where to Look for Innovation

8.2.1. *The structure of the cell*

One structure that is receiving fresh scrutiny is the membrane nanotubes, a thin thread of membrane suspended between cells. It is now known that these structures could span the distance of several cells. Nanotubes could be used as channels to transport small cellular organelles and could also conduct electrical signals, which might enable cells to coordinate during migration or wound healing. Their role in cancer and other diseases needs to be better understood.

When a cell is deprived of purines, the enzymes group together in a cluster called a purinosome. It is postulated that purinosomes can be converted to the cellular fuel adenosine triphosphate and may help power the transport of organelles and materials around the cell on microtubule tracks. Their involvement in the metabolism of proteins provides the nucleotide precursors necessary for mitochondrial adenosine-5′-triphosphate (ATP) production, but also, conversely, they demand ATP for their operation. This metabolic energy source is needed for protein metabolism, and there are many types of protein-forming clumps in starved yeast cells. But it is not clear whether the clumps serve a useful purpose — such as improving metabolic efficiency or acting as storage depots — or are simply a result of cellular failures brought on by starvation. Why is this important? In the case of purines, their levels in mammalian cells are maintained by the coordinated action of complementary salvage and de novo biosynthetic pathways. Although the salvage pathway maintains purine nucleotide levels under normal physiological conditions, the *de novo* pathway is upregulated during growth and altered in neoplastic cells (Antonioli *et al.*, 2013; Yamaoka *et al.*, 2001). Purinosomes are dynamic structures that are formed in response to purine depletion and act to increase de novo purine biosynthesis and their formation in the cells; they are cell cycle dependent (Fang *et al.*, 2013). Therefore, the importance of purinosomes in cancer is that disruption of their formation leads, e.g., to enhanced sensitivity to cancer chemotherapeutic agents. Using super-resolution microscopy, it has been shown that purinosomes colocalized with mitochondria and, interestingly enough, rapamycin (mTOR) influenced the assembly of purinosomes. The inhibition of mTOR reduced purinosomes–mitochondria colocalization and suppressed formation of purinosomes, which is stimulated by mitochondria dysregulation. More data need to be gathered on these structures, an opportunity for innovative research.

Another subcellular package needing additional study is the exosome. Exosomes are tiny membrane-enclosed sacs that form inside the cell and are later excreted. They are present in many, and perhaps all, biological fluids, including blood and urine. Exosomes are released either from the cell when multivesicular bodies fuse with the plasma membrane or directly from the plasma membrane. These structures were discovered in the maturing mammalian reticulocyte and participate in selective removal of many plasma membrane proteins. The plasma membrane is recycled and parts of the membranes of some endosomes are subsequently internalized as smaller vesicles. Such endosomes are called multivesicular bodies, and the intraluminal endosome vesicles become exosomes. Interestingly, exosomes could carry messenger RNA, some of which could be picked up and translated in a recipient cell. This suggests that the shipments might allow cells to affect protein production in their neighbors. The protein content of a single exosome can be about 20,000 proteins. Exosomes have been shown to carry double-stranded DNA. It has also been shown that the ability of the exosomes to transfer molecules from one cell to another via membrane vesicle trafficking may play an important role in regulating the immune system, such as dendritic cells and B cells do, and may even play a functional role in mediating adaptive immune responses to pathogens and tumors (Li *et al.*, 2006). Therefore, the study of exosomes found in body fluids, analyzing their contents to diagnose cancer or deploying exosomes to provoke immune responses against tumors, are other new directions for innovation in cancer research. An interesting area that requires more research is that exosome production and content may be influenced by molecular signals received by the cell of origin, meaning that tumor cells exposed to hypoxia, e.g., could secrete exosomes with enhanced angiogenic and metastatic potential, suggesting that tumor cells adapt to a hypoxic microenvironment by secreting exosomes to stimulate angiogenesis or facilitate metastasis to a more favorable environment (Park *et al.*, 2010). In the case of ovarian cancer, a direct relationship between exosomes release and invasiveness has been shown (Kobayashi, 2014). More importantly, exosomes have been also implicated in the priming of organs for cancer to spread (Kaiser, 2016).

Because exosomes are remarkably stable, their release from tumors into the blood may also have diagnostic potential and could be used for biomarker analysis. However, because the size of an exosome is less than 100 nm, and because they have a low refractive index, they are below the detection range of many currently used techniques. Many methods for the detection of exosomes still need validation and are currently being explored, such as micronuclear magnetic resonance devices, a nanoplasmonic chip, an integrated fluidic cartridge for RNA detection, atomic force microscopy, nanoparticle-tracking analysis, Raman microspectroscopy, tunable resistive pulse sensing, and transmission electron microscopy.

In addition to their potential diagnostic application, the use of exosomes as therapeutic carriers of drugs and micro RNAs — small inference RNA, composed of cellular membranes with multiple adhesive proteins on their surface — could provide an exclusive approach for the delivery of various therapeutic agents to target cells. For example, exosomes are used as a vehicle for the delivery of a cancer drug called paclitaxel. The drug was placed inside exosomes derived from white blood cells, which were then injected into mice with drug-resistant lung cancer; this increased cytotoxicity more than 50 times as a result of nearly complete colocalization of airway-delivered exosomes with lung cancer cells (Batrakova and Kim, 2015; Bell *et al.*, 2016; Kumar *et al.*, 2015; Wahlgren *et al.*, 2016).

8.2.2. *The protein field*

Most protein research focuses on those known before the human genome was mapped. The almost 20 million papers published between 1950 and

2016 are concentrated on three classes of protein, and the same small fraction of family members have remained the favorites for nearly two decades. There is a natural tendency to persist in the known field rather than do research on unstudied proteins, for which it is often harder to explain the rationale and significance. However, if we want to have innovation in cancer research it is important to broaden our horizon in the study and function of protein. DNA microarrays have allowed geneticists to ignore preconceived ideas about disease mechanisms and pursue a remarkably successful broad-brush approach; a similar approach should be embraced for protein studies. This is yet another new frontier that cancer researchers must explore.

8.2.3. *Genomic editing (CRISPR-Cas9)*

CRISPR (clustered regularly interspaced short palindromic repeats) consist basically of a single enzyme called CRISPR-associated protein 9 (Cas9) that can be programmed with CRISPR-derived RNAs to introduce double-stranded DNA breaks at specific sites in the genome. CRISPR-Cas9 was codiscovered by Emmanuelle Charpentier and Jennifer Doudna. The CRISPR–Cas system is the part of the prokaryotic immune system that confers resistance to foreign genetic elements, such as plasmids and phages, thus offering insights and possible treatment alternatives for inherited diseases.

How does CRISPR-Cas9 work? Basically there are two main components of a Cas9 enzyme, one that snips through DNA like a pair of molecular scissors and the other a small RNA molecule that directs the scissors to a specific sequence of DNA where it will make the cut. Then the cell's DNA repair machinery generally repairs the cut.

The CRISPR literature has grown in a logarithmic scale and I strongly recommend an article published by Heidi Ledford in *Nature*, in March 2016, that summarizes the main research trends in CRISPR biology. Among the messages that researchers are interested in unveiling is the specificity, meaning the ability to target and study a particular DNA sequence in the genome. Better managing of Cas9 has recently been attained (Kleinstiver *et al.*, 2016; Slaymaker *et al.*, 2016), which allows the enzyme to retain its capacity to cut DNA in an RNA-guided manner but deletes the undesirable properties if the enzyme is cutting multiple unintended sequences. The researcher-engineered Cas9, paired to recognize and cut its target, has been almost 75% more efficient in targeting the cut than the wild form of Cas9. The advantage of this new and improved Cas9 is that it will reduce the time it takes to complete a genome-editing experiment because the need to check for undesired edits will be reduced. The implication of this finding is enormous because when Cas9 was brought in from the wild and placed in human cells, it introduced genetic changes to unintended stretches of DNA in addition to editing the gene of interest (Tsai *et al.*, 2015). Additional studies at the molecular levels of CRISPR-Cas9 will advance the field even farther and speed up the clinical application of gene editing.

As discussed by Ledford (2016), there are several ways to control CRISPR, one of them involves using "broken scissors", another controls CRISPR epigenetics by producing an inducible CRISPR. The broken scissors method makes Cas9 nonfunctional through mutation, meaning that it is unable to cut the DNA but still with the right guide RNA attached to a specific part of the genome. The dead Cas9 blocks the binding of other proteins, like RNA polymerase that is needed to express the gene. The beauty of this idea is that an activating protein can be attached to the dead Cas9 protein to stimulate expression of a specific gene. Using the same principle, the broken Cas9 enzyme can be coupled to epigenetic modifiers, e.g., adding methyl or acetyl group to histones and in that way have a better understanding of the mechanism of gene expression. Dead or alive, the Cas9 can also be coupled to switches, which can be controlled by chemicals or by light, for example.

A preclinical model of diseases is essential for starting clinical trials. Therefore, the engineering of animals using CRISPR-Cas9 will allow the editing of the genome to induce specific types of tumors by changing the genomic composition. Mimicking this tumor is the next frontier in cancer research.

The use of CRISPR-Cas9 has raised some concerns for its function in performing germ-line genetic modifications, which means making changes in a human egg, sperm, or embryo, creating modifications that will be passed down to generations, impacting an entire lineage, rather than just one person. However, among the various demonstrated applications is that Cas9 protein increases the amount of utrophin, a known disease-ameliorating protein in Duchenne muscular dystrophy (DMD). In addition, removal of a duplication of DMD exons 18–30 in myotubes of an individual affected by DMD produced full-length dystrophin, ameliorating the effect of the disease. Guidelines for the use of this technique were established by a committee at the end of 2015. The committee was against the use of gene editing in regards to the germ line.

Finally, we must acknowledge that the positive and innovative aspect of CRISPR-Cas9 research is that it provides an alternative approach to gene editing that can be directed toward altering only somatic cells and is therefore a new and invaluable tool for cancer research.

8.2.4. *Long noncoding RNA (lncRNA)*

A noncoding RNA (ncRNA) is a functional RNA molecule that is not translated into a protein. The DNA sequence from which a noncoding RNA is transcribed as the end product is often called an RNA gene, or noncoding RNA gene. Noncoding RNA genes include highly abundant and functionally important RNAs such as transfer RNA (tRNA) and ribosomal RNA (rRNA), as well as RNAs such as snoRNAs, microRNAs, siRNAs, and piRNAs, and the long ncRNAs that include examples such as Xist and HOTAIR. The number of ncRNAs encoded within the human genome is unknown; however, recent transcriptomic and bioinformatics studies suggest the existence of thousands of ncRNAs. Since most of the newly identified ncRNAs have not been validated for their function, it is possible that many are nonfunctional but a significant number of them may have a crucial role in regulating gene transcription.

Noncoding RNA is generated in eukaryotes by the spliceosomes that perform the splicing reactions essential for removing intron sequences; this process is required for the formation of mature mRNA. Another group of introns can catalyze their own removal from host transcripts; these are called self-splicing RNAs. There are two main groups of self-splicing RNAs: group I, catalytic intron, and group II, catalytic intron. These ncRNAs catalyze their own excision from mRNA, tRNA, and rRNA precursors in a wide range of organisms. These ncRNA are involved in critical regulatory processes that define biological processes and roles of a cell. The expression of these molecules has been shown to be a systematic transcriptional and regulatory event involving recruitment of splice-osomes machinery and nuclear export. The majority of these ncRNA are expressed during various stages of development and are highly tissue specific. It is understood that these molecules exert a higher tier of regulation over critical structural and regulatory proteins and are essential to epigenetic processes and directing cell fate.

The results from the initial transcriptomic analysis in the normal breast of parous and nulliparous women reveal that genes such as nuclear paraspeckle assembly transcript 1 (*NEAT1*), MALAT-1 (*NEAT2*), and X-inactive specific transcript (*XIST*) are upregulated in the parous breast (Belitskaya-Lévy *et al.*, 2011; Peri *et al.*, 2012; Russo *et al.*, 2012). The role of these genes in epithelial cell differentiation and maturation needs to be explored. More importantly, we must assess the role of ncRNA and the genes involved in epithelial cell maturation and fate in both parous and nulliparous women. Furthermore, the expression levels of these molecules can be combined with expression measurements of coding RNA to identify the systematic regulatory networks that are involved in parous and nulliparous breasts and could provide novel and significant information in the role of ncRNA in breast cancer prevention.

To emphasize the importance of long noncoding RNA, I would like to give an example. The development of the organs of the human body is a lifelong process of organogenesis that starts in the embryo; female organs are established at the time of fertilization of the oocyte through the inheritance of a paternal X chromosome. Subsequent cell divisions will generate 2, 4, and 8 cells, which will progress to the morula stage (human pre-embryo) and the blastocyst, within which the inner cell mass (ICM) cells continue to proliferate and constitute the ultimate totipotent cells from which the embryonic stem (ES) cells are harvested. The ICM cells represent the ultimate undifferentiated cell type, as it gives rise to all cell types and live offspring. Totipotency persists for the very first cell doublings, from the single cell and zygote to at least the 4-cell pre-embryo. Initiation of transcription in the newly formed embryonic genome reportedly occurs at the 4- and 8-cell stage, followed by decreases in abundance of individual mRNAs. The XX female needs to equalize her gene dosage relative to XY males by inactivating one of her X chromosomes. An inactive chromosome X (Xi) becomes present in each cell through the X chromosome inactivation center (XCI), which contains several genetic elements essential for the transcription initiation of long noncoding RNAs (lncRNAs). Initiation of XCI requires the accumulation of a nontranslated human XIST RNA that coats the X chromosome and is followed by various epigenetic changes on the future Xi that contribute to chromosome silencing, which is detected as early as the 4-cell stage of mouse embryos. Defects in dosage compensation before implantation lead to abnormal development in a majority of the embryos and early lethality. Thus, the presence and function of lncRNA determines the lifetime ability of the breast, e.g., to develop normal growth and differentiation, which require a synchronized sequence of cellular processes that involve commitment to specific cell types and its ultimate function, milk production. Any derangements affecting the integrated framework of signaling networks that control breast development during puberty, pregnancy, lactation, or post-lactation involution, such as inheritance or exposure to genetic and environmental factors that cause accumulation of germ line or epigenetic mutations, might lead to the neoplastic transformation. Loss of Xi has been reported in both BRCA1-defective and in wild breast cancer cells. XIST is the key player of the X chromosome silencing; the inactive X (Xi) acquires typical features of heterochromatin, such as late replication, hypoacetylation of histones H3 and H4, methylation of histone H3 at lysines 9 and 27, lack of methylation of H3 at lysine 4, and methylation of DNA CpG islands, epigenetic modifications that appear to act synergistically and stably maintain the inactive state through subsequent cell divisions. Studies of human embryonic stem cells (hESCs) have demonstrated that lncRNAs are important regulators of pluripotency and neurogenesis, playing an important role in human brain development. Similarly, the role played by lncRNAs, and more specifically XIST, in normal development and differentiation of the breast has been confirmed by our studies of the genomic and epigenetic characteristics of the differentiated breast of early parous women who had reached menopause free of breast cancer (Russo and Russo, 2013).

Overall the importance of the lncRNA in cancer cannot be emphasized enough; there is no doubt that this new frontier must be explored.

8.2.5. *Splicing mechanism*

In 1977, work by the Phillip Sharp and Richard J. Roberts laboratories revealed that genes of higher organisms are "split" or present in several distinct segments along the DNA molecule. The coding regions of the gene are separated by noncoding DNA that is not involved in protein expression. The introns are excised from the precursor mRNAs in a process Sharp named "splicing". The advances in this field are unprecedented, providing a better understanding of the basic biological process and its application to the treatment and prevention of many diseases (Russoand Russo, 2013).

Basically, a spliceosome is a complex of specialized RNA and protein subunits that removes introns from a transcribed pre-mRNA (hnRNA) segment. This process is generally referred to as splicing. Each spliceosome is composed of five small nuclear RNA proteins, called snRNPs, and a range of non-snRNP associated protein factors. The snRNPs that make up the nuclear spliceosome are named U1, U2, U4, U5 and U6, and participate in several RNA–RNA and RNA–protein interactions. The canonical assembly of the spliceosome comprises the conversion of precursor messenger RNA into mature messenger RNA (mRNA). The pre-mRNA molecule undergoes three main modifications, which occur in the cell nucleus before the RNA is translated: (1) mRNA capping, (2) processing of intron-containing pre-mRNA, and (3) mRNA 3′-end processing. These three main steps are executed by a large number of proteins that we have discussed in detail in our book *Role of the Transcriptome in Breast Cancer Prevention* (Russo and Russo, 2013). Recently a more detailed molecular-resolution reconstruction of a central assembly of the human spliceosome — the U4/U6.U5 triple small nuclear ribonucleoprotein (tri-snRNP) complex, using cryo-electron microscopy (cryo-EM) and the X-ray crystal structure of the U1 snRNP — provides structural models of the splicing machinery, launching a new era in understanding eukaryotic gene regulation (Agafanov *et al.*, 2016).

More research will afford a better understanding of the splicing mechanism that can provide new clues on how a single gene has the potential to encode many different protein variants with diverse properties and functions. This is definitely a new area that cancer researchers must continue to explore for a better understanding of the prevention and treatment of cancer (Russo and Russo, 2013).

8.3. Creative Science: Thinking as a Path to Innovation

Of all the cognitive learning skills, creative thinking is the most complex and abstract. **Creative thinking allows the restructuring of a problem and producing solutions using new insights.** Basically, creativity is the root of innovative thinking; it leads to solutions or products that are novel, useful, and critical to economic success. In science, we are dealing with experiments and analyses of data that lead us to conclusions, and from there we elaborate hypotheses and new ways of thinking that may or may not lead us to an innovation. In science, there is a vague notion that creative minds are in the areas of music and arts, such as painting or literature; however, the use of creative thinking can bolster the dimension of data analysis leading to innovation — that is a definitive way to see things.

It has been shown that about one-fourth of US college students have the reasoning skills necessary to solve conceptual problems (DeHaan, 2011); that is probably because creativity is a complex, multicomponent construct and, therefore, it is not easy to define or assess, especially in the context of science. It is required to produce creative insight in an individual mind as well as associative thinking, in which intuition may not only play a role, but also increases the probability of connecting weakly associated ideas. A second important component of an innovative idea is analytical thinking, i.e., the capacity to analyze, synthesize, and focus on specific sets of data or results obtained by observations or empirically. As has been elegantly analyzed in a publication of DeHaan (2013), new research needs to emphasize the roles of both associative thinking and distributed-reasoning play in science and in creativity; this leads to innovation. But fostering associative thinking and peer-to-peer interactions during the problem-solving discussion can improve the originality and novelty of the solutions to the problems that we have in cancer research.

In an article by Leshner (2011), he emphasized that innovation demands that novel ideas be pursued, and indicated that the advancement of science and technological enterprise may not be well structured in places other than the United States. He also points out that in the

United States, peer review can be somewhat conservative, particularly when grant support is limited, resulting in the support of "safer" projects over riskier ones that are based on scant pilot evidence (Leshner, 2011). This shortness of vision is jeopardizing the innovative edge that is needed to succeed in long term. Although some funds are oriented in this direction, the number of funded projects is limited, forcing many innovative ideas to merge. Therefore, the new cadre of cancer researchers must embrace this paradigm in which innovation is the result of creative minds.

8.4. Motivation by Leadership of the New Generation of Cancer Researchers that Leads to Innovation

If we aim for innovation in cancer research we must start by motivating each other. The bottom line of motivation is how to get the most out of the people or team that you work with without overwhelming them or micromanaging their activities. The difficulty in motivating a research team is that, in general, they are composed of a heterogeneous group of individuals. Each of them could be smart and self-determined individuals, such as junior researchers, research associates, postdocs, graduate students, and research assistants, but all of them are individuals who have different goals. This heterogeneous group of individuals must be led to succeed using their own initiative, drive, and creativity. The most important thing in choosing members of a research team is to be sure that they understand the overall aim of the research laboratory. This is probably the most important aspect of the research team. If one member is interested in working on an idea that is not the leading one of the group, that person should probably not be part of the team, or at least the leader of the group should use this specifically motivated person to develop a sub aim of the project that can contribute to the group's main aim. Although it is difficult to suggest a specific recipe that applies to everyone, the wisdom of the team leader is to capture the individual interests and capabilities of his or her team members. Using this basic principle to find the passion, or driving force, of each individual is probably the most important skill of a wise leader. This will allow every member to focus either individually or as a part of a small team on the sub aims of the main aim of the research project. Discovering the passion of each individual will help the leader of the group maintain collective interest in the project's focus and foresee the impact of the work on the big picture of science. This is also a recipe for avoiding frustration and being able to wisely evaluate obstacles to the research at hand.

If a research leader wants to accomplish innovation in their research endeavor, he or she must also learn to deal with different personality types. Some members of the group might be perfectionists, even to the point that they are becoming an obstacle to accomplishing the final goal; other members might be able to see only the big picture but be unable to see the intermediate steps needed to accomplish the main goals. Although these are two extremes of the same line, it requires the wisdom of the team leader to deal with these two different personality profiles. Wisdom of the team leader implies also having the flexibility to not only praise small and great achievements of team members but also to recognize that mistakes can be the path to future success. As I have indicated, the leader must motivate his team to creative thinking and see not just the problem from different perspectives and but also how to provide a solution to that same problem. In my book, *The Tools of Science* (Russo, 2010), I have addressed the importance of critics and how they can be used to help the team achieve success, mainly as motivators for innovation. The basic rule is not to be afraid of critics, rather welcome them as a renewal force for success and creativity.

It is also important that the team leader understands his or her limitations as well as the human limitations of the members of his or her team. Some human limitations in knowledge, technical skills, e.g., can be corrected; however, limitations imposed by the individual burden, physical, mental, or societal, must also be dealt with but they are not part of this essay.

Finally it is important to accept that there will be a small group of very talented people that stimulate breakthrough and innovation. *Doing innovative things requires smart people with vision and courage who will focus on the right things.*

Another important concept that must permeate science and biomedical research, in particular, is that **biology is difficult,** and it requires time to acquire the wisdom that is needed to break new frontiers. In contrast, mathematics and physics or even other nonbiological sciences are also complex but more predictable than the study of biological systems. That is the reason why technology has advanced at a faster pace; they are dealing with constants and predictable outcomes. The complexity of living systems makes it difficult to establish a priority in the research endeavor, as it is more complex than the sum of different disciplines associated with living science. The best example is cancer research. As I have described in Chap. 1, the choice of which subject to investigate opens the career of the cancer researcher. This requires a large understanding of the biological and pathological processes and makes the choice especially challenging for the new cadre of cancer researchers. This warning is also for mentors, the ones who are orienting the study subject of the young cancer researchers.

8.5. Innovation through Partnership

Universities and independent research institutes can find allies in the business world. In the United States, this partnership has been very well nurtured through the years. When this partnership is established, the benefit is not only the monetary reward for academia but also the practical implications of the collaboration. Among them is the **need** to have more immediate societal or economic impacts. This partnership can easily translate solutions to a larger sector of the population by bringing an idea into real-world practice and turning a research breakthrough into a global application — and that is true innovation. The need for innovative therapeutics to overcome intractable diseases in immunology, such as cancer, requires integration of clinical science in drug discovery and development. For this purpose, a real interdisciplinary synergy among basic scientists, clinicians, and scientists is needed. This requires sharing clinical information and samples, knowledge of molecular mechanisms, and drug discovery skills. Pharmaceutical companies can establish collaboration with research institutions for drug discovery and development in cancer treatment and prevention. The pharma industry can provide various technology platforms to analyze the highly accurate patient information and clinical samples available through the university's medical school and hospital, linking the clarification of the basic structure and the mechanism of living organisms.

Phase I and II clinical trials for proof of concept at the translational research center scan should be a real partnership between academia and companies: scientific inputs, outputs, knowledge, and intellectual properties can be shared with both parties. However, these models of collaboration are not easy to accomplish in part due to the complexity of cancer, which requires finding molecular patterns of behavior that need to be obtained from the tremendous amount of data pouring out of big genomics projects. The analysis of these genomics databases could provide the intracellular pathways that contribute to cancer and facilitate building computational models of those conditions and potentially using them in preclinical drug development. The new cadre of cancer researchers must embrace this partnership and be very inclusive in their way of thinking and working.

8.6. The Sustainability of the Cancer Researchers and the Development of Innovative Science

The premise is that the individual scientist maintains independent research endeavor and is the driving force determined by his or her capability; that, in turn, must be the source of innovation and the seed for translational research. This premise is a good one and the truth is that it has been the experience or the natural process of many of us as cancer researchers. Based on this premise, cancer research endeavors have flourished and magnified exponentially during the last 40 years. However it has been based on the assumption that the government must pay for this; therefore, the main element of this drama is that a researcher's sustainability is not guaranteed, it depends on the ability of the researcher to compete. This competition has created the quality of science that all of us are proud of. A very astute analysis by Alberts (2010) led us to reflect on the weakness of this equation. He said "that a reliance on the NIH to pay not only the salaries of scientists but also the overhead (or indirect) costs of building construction and maintenance has become a way of life at many US research institutions, with potential painful consequences". One consequence has been encouraging American universities, medical centers, and other research institutions to expand their research capacities indefinitely through funds derived from the National Institutes of Health (NIH), as research grants make the situation unsustainable.

This dependency on the NIH is so dramatic that the sustainability of the researchers is the factor that determines the indirect cost that the institutions receive funding for, and as a consequence, there is no security for the cancer researchers to maintain their positions unless this extramural support is maintained. The real problem is that the NIH budget cannot increase at high enough rates to pay for the ever-expanding US biomedical research enterprise. Here is where the system fails because the resource (money) cannot grow at the same speed as the number of researchers in the field, and thus the competition determined by the scarcity of funds keeps the pay line of the grants extremely low (around 10%), making the situation a desperate one for cancer researchers. The concepts of innovation, impact, and significance are required when competing for this small piece of the pie. Research and academic institutions depend on the individual investigator to maintain the research endeavor, and this requires that the investigator be self-sufficient in a certain way, in order for the institution to guarantee a salaried position. To maintain their position in this race, researchers need to spend a significant part of their time writing grant applications that have less than a 10% chance of succeeding. It is in this environment that the concept of innovative science must flourish.

The problem is a complex one and it is the appropriate time for academia and research institutions, together with the NIH, to do some real soul-searching on how to maintain a driving force of creative, motivated researchers and provide an environment of healthy competition but at the same time cut the Gordian knot that is the sustainability of the cancer researchers. With the scarcity of funds for research, research-originated grants by individual researchers, or RO1, are competing with other sources of grants, such as program projects, core grants, and SPORE, making the innovative investigator-initiated research suffer. If emphasis is put on innovation, the curtailing of RO1 makes the future of medical progress more uncertain. Why is it so important to maintain the RO1? Because they are the primary source of new knowledge and discoveries. In the perception of Rosbash (2011) "the most important breakthroughs often come from unexpected areas of inquiry. For example, recombinant DNA and monoclonal antibodies emerged from fundamental research in bacteriology and immunology, respectively. These technologies gave birth to the biotechnology industry and underlie many therapeutics approved by the U.S. Food and Drug Administration" (Rosbash, 2011).

I cannot provide a solution to this problem of the sustainability of cancer researchers because it is not an easy equation to solve. Many different components are in play; economic and social pressures and national priorities are all involved in different dynamic proportions. I am not the first to point to the problems and it will require significant brainpower from the new cadre of cancer researchers to find a viable path.

References

Agafonov, D.E., Kastner, B., Dybkov, O., Hofele, R.V., Liu W.T., Urlaub, H., Lührmann, R., and Stark, H. Molecular architecture of the human U4/U6.U5 tri-snRNP. *Science*, **351**: 1416–1420, 2016.

Alberts, B. Overbuilding research capacity. *Science*, **329**: 1257, 2010.

Antonioli, L., Blandizzi, C., Pacher, P., and Hasko, G. Immunity, inflammation and cancer: A leading role for adenosine. *Nat. Rev. Cancer.* **13**: 842–857, 2013.

Batrakova, E.V. and Kim, M.S. Using exosomes, naturally-equipped nanocarriers, for drug delivery. *J. Control Release.* **219**: 396–405, 2015.

Belitskaya-Lévy, I., Zeleniuch-Jacquotte, A., Russo, J., Russo, I.H., Bordas, P., Ahman, J., Afanasyeva, Y., Johansson, R., Lenner, P., Li, X., Lopez de Cicco, R., Peri, S., Ross, E., Russo, P.A., Santucci-Pereira, J., Sheriff, F.S., Slifker, M., Hallmans, G., Toniolo, P., and Arslan, A.A. Characterization of a genomic signature of pregnancy identified in the breast. *Cancer Prev. Res.* **4**: 1457–1464, 2011.

Bell, B.M., Kirk, I.D., Hiltbrunner, S., Gabrielsson, S., and Bultema, J.J. Designer exosomes as next-generation cancer immunotherapy. *Nanomedicine*, **12**: 163–169, 2016.

DeHaan, R.L. Teaching creative science thinking. *Science*, **334**: 1499–1500, 2011.

Fang, Y., French, J., Zhao, H., and Benkovic, S. G-protein-coupled receptor regulation of de novo purine biosynthesis: A novel druggable mechanism. *Biotechnol. Genet. Eng. Rev.* **29**: 31–48, 2013.

Kaiser, J. Malignant messengers. *Science*, **352**: 164–166, 2016.

Kleinstiverl, B.P., Pattanayakl, V., Prew, M.S., Tsai, S.Q., Nhu, T., and Nguyen, N. High- fidelity CRISPR-Cas9 nucleases with no detectable genome-wide off-target effects. *Nature.* **529**: 490–495, 2016.

Kobayashi, M. Ovarian cancer cell invasiveness is associated with discordant exosomal sequestration of Let-7 miRNA and miR-200. *J. Transl. Med.* **12**: 4, 2014.

Kumar, L., Verma, S., Vaidya, B., and Gupta, V. Exosomes: Natural carriers for siRNA delivery. *Curr. Pharm. Des.* 21: 4556–4565, 2015.

Ledford, H. Riding the CRISPR wave. *Nature*, **531**: 156–159, 2016.

Leshner, A.I. Innovation needs novel thinking. *Science*, **332**: 1009, 2011.

Li, X.B., Zhang, Z.R., Schluesener, H.J., and Xu, S.Q. Role of exosomes in immune regulation. *J. Cell. Mol. Med.* 10: 364–375, 2006.

Park, J.E., Tan, H.S., Datta, A., Lai, R.C., Zhang, H., Meng, W., Lim, S.-K., and Sze, S.K. Hypoxic tumor cell modulates its microenvironment to enhance angiogenic and metastatic potential by secretion of proteins and exosomes. *Mol. Cell Proteomics.* **9**: 1085–1099, 2010.

Peri, S., López de Cicco, R., Santucci-Pereira, J., Slifker, M., Ross, E.A., Russo, I.H., Russo, P.A., Arslan, A.A., Belitskaya-Lévy, I., Zeleniuch-Jacquotte, A., Bordas, P., Lenner, P., Åhman, J., Afanasyeva, Y., Johansson, R., Sheriff, F., Hallmans, G., Toniolo, P., and Russo, J. Defining the genomic signature of the parous breast. *BMC Med. Genomics.* **5**: 46, 2012.

Rosbash, M.A. Threat to medical innovation. *Science*, **333**: 136, 2011.

Russo, J. *The Apprentice of Science: A Handbook for the Budding Biomedical Researchers.* World Scientific Singapore, 2010.

Russo, J. and Russo, I.H. *Role of the Transcriptome in Breast Cancer Prevention.* Springer, New York, 2013.

Russo, J., Santucci-Pereira, J., López de Cicco, R., Sheriff, F., Russo, P.A., Peri, S., Slifker, M., Ross, E., Mello, M.L.S., Vidal, B.C., Belitskaya-Lévy, I., Arslan, A., Zeleniuch-Jacquotte, A., Bordas, P., Lenner, P., Ahman, J., Afanasyeva, Y., Hallmans, G., Toniolo, P., and Russo, I.H. Pregnancy-induced chromatin remodeling in the breast of postmenopausal women. *Int. J. Cancer.* **131**: 1059–1070, 2012.

Slaymaker, I.M., Gao, L., Zetsche, B., Scott, D.A., Yan, W.X., and Zhang, F. Rationally engineered Cas9 nucleases with improved specificity. *Science*, **351**: 84–88, 2016.

Tsai, S.Q., Zheng, Z., Nguyen, N.T., Liebers, M., Topkar, V.V., Thapar, V., Wyvekens, N., Khayter, C., Iafrate, A.J., Le, L.P., Aryee, M.J., and Joung, J.K. GUIDE-seq enables genome-wide profiling of off-target cleavage by CRISPR-Cas nucleases. *Nat. Biotechnol.* 33: 187–197, 2015.

Yamaoka, T., Yano, M., Kondo, M., Sasaki, H., Hino, S., Katashima, R., Moritani, M., and Itakura, M. Feedback inhibition of amidophosphoribosyltransferase regulates the rate of cell growth via purine nucleotide, DNA, and protein syntheses. *J. Biol. Chem.* **276**: 21285–21291, 2001.

Wahlgren, J., Statello, L., Skogberg, G., Telemo, E., and Valadi, H. Delivery of small interfering RNAs to cells via exosomes. *Methods Mol. Biol.* **1364**: 105–125, 2016.

Further Reading

Dunbar, K. In *Mechanisms of Insight*, R.L. Sternberg, L. Davidson (eds.). pp. 365–395. Cambridge: MIT Press, 1995.

French, J.B., Jones, S.A., Deng, H., Pedley, A.M., Kim, D., Yu, C., Hu, H., Pugh, R.J., Zhao, H., Zhang, Y., Huang, T.J., Fang, Y., Zhuang, X., and Benkovic, S. Spatial colocalization and functional link of purinosomes with mitochondria. *Science*, **351**: 733–737, 2016.

National Science Board, Preparing the next generation of STEM innovators: Identifying and developing our nation's human capital. www.nsf.gov/nsb/ publication 1033, 2010.

Chapter 9

The Past and Present of Academic Research

9.1. The Basic Concept of the Old Academic Environment

In the early 1950s, the concept of "publish or perish" was created as a gatekeeper to separate sound research from shoddy or biased counterparts. The concept of publish or perish was the imperative to constantly publish work to further or sustain an academic career. Although there were severe critics of this concept, the number of publications was a convenient metric to measure faculty productivity in academia, and this also generated the need for multiauthorships; thus the data were divided into small aliquots in order to publish one idea or experiment per paper. Doctoral theses were parceled in small portions and published even before the doctoral thesis was defended. This game of numbers has been imitated faithfully in China and India, as well as in Europe and the rest of the Americas, and it is a common metric to determine the scientific productivity of a country's academic endeavor. As a way to counteract quantity for quality, other sets of metric were established, like impact factor of the journal or bibliometric indices, such as the h- and g-indexes, based on citation counts, to evaluate a researcher's impact in their discipline (Russo, 2010). However, these measures did not decrease the number of papers, but on the contrary caused the inflation of self-citation as well as the citing of friends who then cite in return.

This "publish or perish" notion has been paralleled by the increase of journals and the need for reviewers, making the problem even more complex. The highly skilled scientists with significant prestige and knowledge are too busy to spend time reviewing papers, and as a consequence the reviewing, or peer review, process is less strict and less balanced than it was 20 or 30 years ago. This concept applies to the peer review system of grant applications, which is in the hands of either junior faculty or not-well-seasoned scientists making the process difficult to control (Russo, 2010).

The future of academic research will depend on sticking to the main objective of science: namely, creating and distributing new knowledge, and for that a high sense of scholarship must be implemented. The publications generated must aim to produce a solution to a problem and the peer review must be maintained at a high standard by the reviewers and publishers alike. The question is how many institutions around the world will be able to keep these standards?

Until around 1980 many African nations lacked infrastructure, China was isolated, India mired in poverty, and South Korea devastated by war and occupation; things changed when the leaders of some of those nations, like South Korea, Taiwan, and then China, began to invest in science — their futures were linked to R&D. In India, the contrast between the large cities and the rural areas intensified. There are around 90 million Indians between the college-going ages of 17 and 21, rising to an estimated 150 million by 2025; these kids are hungry and even starving for an education (Nayar, 2011).

In the publication of Anjali Nayar is well addressed how the population growth, has fueled an eightfold increase in science and engineering enrollment at India's colleges and universities over the past decade; unfortunately, the vast majority of India's science and technology graduates immediately head for high-paying jobs in industry and only about 1% of them go on to get PhDs, compared with about 8% in the United States. All the talent goes into sectors that make money but produce very little in terms of creative things for the country. The most depressing aspect is that the 15 Indian Institute of Technology campuses nationwide have roughly 300,000 applicants every year and only 2% are accepted, meaning that there is really a significant gap between the demand for education and the places that offer it. According to Anjali Nayar, India will need another 800–900 universities and 40,000–45,000 colleges within the next 10 years. How this will be solved is anybody's guess.

9.2. The New Emphasis in Academic Research

There are significant new topics that must be emphasized in the training of cancer researchers that it will be beyond the scope of this book; therefore, I have selected those topics that in my experience are of relevance.

9.2.1. *Scientific writing*

Scientific writing is best learned in the context of doing science. Because students "do" science (as opposed to "learn about" science) almost exclusively in laboratory courses, they need to learn the skills of scientific writing there, according to Moskovitz and Kellogg (2011) publication in *Science*. Scientific writing involves a variety of rhetorical functions, including persuading skeptical audiences, constructing interpretive frameworks, refuting the work of others, and so forth. Therefore, these topics need to be part of the curriculum of training for our next generation of cancer researchers. The key skills in communicating science are selecting which data to present and learning forms of writing that working scientists use.

My position on scientific writing is that it is a craft that goes hand in hand with the making of science because it is the process of writing that makes the ideas and interpretations of our own discoveries visible to others. This craft is the one that needs to be emphasized by our educational institutions and by our mentors. There is a tendency and a temptation to use the available manuscript-editing marketplace, as discussed by Perkel (2016), as a good alternative for non-native English speakers or for those who prefer the writing be delegated work that you can pay for. At the end, there is nothing wrong with doing this because it adds clarity to the written words and there are good skillful writers who can put ideas together better than the individuals who originated the ideas. However, I need to make the point that it is important for cancer researchers to craft the writing because a manuscript is a permanent record or memory of the life expended by that individual in creating the new knowledge that will stay forever. Therefore, good writing is part of the creative process of science.

9.2.2. *Scientific ethics*

Because there is a significant incremental knowledge of new technologies that can be applied to human health, like stem cell research or nanotechnology, there is an important ethical question raised by the generation of young researchers: *How ought we communicate about the promise of novel biotechnologies with the aim of catalyzing public support while avoiding hype?* And probably the best answer is to request that the researchers and the media provide accurate reporting, accounting for the scientific merits and caveats of research by showing not only the progress but also the road blocks and consequences of the nature of the research in question.

Another challenging ethical question facing young investigators is a lack of scientific integrity, which may damage the whole scientific enterprise. Although there are multiple safeguards to stop this process, such as education, ethics review boards, and healthy research environments, probably the simplest concept that must permeate the minds of the present as well as future generations of researchers is that to do research is an option and that option must be freely chosen by the individual. As I already described in my book *The Apprentice of Science: A Handbook for the Budding Biomedical Researchers* (Russo, 2010):

> the scientific life is unlike any other. Scientific research is not simply a 'job,' rather a calling to discover the beauty of truth. Besides the elements of curiosity and an urge for understanding, which are the characteristics of a scientist, one element that on many occasions is not very well considered or described is the idea of vocation. Vocation, meaning "a summons or strong inclination to a particular state or course of action," stems from the Latin word vocare, which means "to be called." While most of the time this word is used in its religious sense, "a divine call to the religious life," as in answering the call to priesthood, I think this concept also applies to those doing cancer research. An individual may have adequate intellectual acumen, or IQ, and be full of curiosity about nature's barest mechanics, but not be truly fit to embark on the journey of scientific discovery. To be a scientist requires commitment and a disciplined mind because good science that will in the end be useful for other human beings requires time, skill, and patience. Therefore, I feel that personal motivation, or a sense of vocation, should be an important factor in considering a life in the scientific field. I want to

emphasize that the commitment required to pursue a career in science, while lofty, should not be easily discarded. Many years of training, understanding the basic sciences and the knowledge that needs to be communicated and utilized, are important to science as a whole, and this needs to be imprinted in the minds of those who feel they are called to be scientific researchers.

If a research scientist pursues his or her endeavor in this manner, I will not worry of plagiarism or misconduct.

The use of animals in scientific research is an ethically challenging issue facing many young researchers. The problem is rooted in a misunderstanding by the public of the importance of animals in research and also, in a certain way, stimulated by the media. The best way to overcome this is by constant public engagement, social communication, and early education of people, explaining why the laboratory animals are indispensable.

The complexity of medicine and the new genetic tools available have created an ethical dilemma for researchers: how to communicate with the source of a sample under a genetic test that is only part of a multiple sample study, or how to explain to the patient complicated sets of data coming from DNA sequencing or the possibility to use genome editing to cure cancer. These are genuine questions that are a burden for the present generation of cancer researchers. All of this emphasizes the need to implement formal training for our new generation of scientists and clinicians about these aspects of knowledge translation, as well as educating the public about medical tests and procedures, so that patients can truly become empowered to make informed decisions.

Study of environmental issues and their effects on human health, mainly cancer, are only partially developed, and, whereas branches of the NIH are interested in pursuing research, there are not enough funds to cover the tremendous amount needed to solve critical questions on the role of environmental exposure in young age, puberty, and adulthood. The need to understand, e.g., the windows of exposure to susceptibility to cancer is a critical area to be studied, mainly how the effect of exposure at transgenerational levels can affect the genomic and epigenetic composition of the human tissues to develop cancer.

The security of data has never been more vulnerable than now, not only the challenge of protecting human data confidentiality but also guarding against the unscrupulous misuse of databases. Ensuring secure databases is beyond the regulations on accessing data because the data must be inaccessible to hackers and pirates, from both internal and external sources that were not intended to have access to them.

9.2.3. *Emulation as a first step in scientific inquiry*

The ability to learn from others is central to the evolution and persistence of culture and it is viewed as part of the reason humans have come to dominate the planet (Pennisi, 2010). It is common sense to watch and imitate what others do in an unknown environment, and could even be a survival mechanism. The vital question is whom do we follow and imitate and learn from. As small children we learn from our parents and relatives to do simple tasks at home, like work in the yard or make our beds. The same applies in laboratory work; watching how another scientist pipets, for example, helps us to emulate and start our motor dexterity. At a more complex level, copying how others think and work in a scientific endeavor helps to establish some pattern of thinking. It is a known fact that taking apart an instrument helps engineers to construct the same device. It is true that this method of copying is not the orthodox way to do things but emulating the behavior of another, like how to be in a scientific environment, helps the novice learn how certain things are done. Shadowing a scientist allows a young person to experience laboratory life and procedures, and imitating and emulating could be a good starting point — those that have been trained in the medical arts know how important it is to see other doctors do their work. The message for the present cadre of cancer researchers is to learn how things are done by observing and emulating seasoned cancer researchers and the way they handle problems, personnel, budgeting decisions, and also personal interactions.

We as humans have a tremendous ability to assimilate what we see and shadowing is an important learning tool that we must not only learn but also teach others to use proficiently. The connections between mentor and mentee, teacher and student, parent and child are all combinations of the same pattern of learning that is not **the mere acquisition of knowledge but the affective part of our human relations, which makes the knowledge acquired more appreciated.** I will never forget when my first science mentor showed me how to handle a mouse and how to put it to sleep and where to look for specific lesions; emulation of these cherished moments has been a tremendous inspiration for teaching others how to be scientists. It is true that these days we can find almost every single thing that we want to learn on the Internet and it is tempting to deny the importance of the mentor–mentee relationship, but **my vision is that even more important than the knowledge we receive shadowing experts are the human emotions and warm feelings we get from working with other humans.**

As I finished writing this chapter, Rumsey (2016)'s book *When We Are No More*, which discusses the future of memory in the digital era, was called to my attention. The interesting part of this work is showing how much we rely on the digital data and how little on our memories. I personally wonder how all of this will affect our ability to achieve more advanced, abstract thinking. In a normal conversation, people frequently refer to something that they read on the Web but when they want to repeat the information, they do not remember very well and they need to go back to their smartphone to retrieve or complete the idea. Rumsey's book also brings our attention to our readiness to discard the written words in books or ancient documents that were treasured for centuries and now are less important because we have access to their contents through digital repositories. Although all these advances can without doubt help people access knowledge, in my perspective they can also ameliorate our ability to remember or to exercise our memories. We must continue to treasure the accumulated works in books and remember the human importance of touching a book that has been written by another human being. This is another human perspective that I emphasize because it helps us to respect intellectual property when we understand that a specific person has created and recorded that knowledge. The anonymity of the digital era can make us wiser if we choose to use it properly and don't become less attentive to the efforts of others.

9.2.4. *The value of communications and conferences in the formation of cancer researchers*

Conferences can be a great trigger for early career scientists, offering countless opportunities to meet mentors and collaborators. Preparing to speak at conferences could be learned by trial and error, but there is also wisdom to learn by emulating good speakers. In my book *The Apprentice of Science: A handbook for the Budding Biomedical Researchers"* (Russo, 2010), I advocated that:

> from early on the scientific apprentice is called to the podium to discuss papers, or present preliminary or final data to be discussed by and exposed to their peers. Unless a scientific apprentice has experience on a debate team or something of the like, this kind of public speaking [is new ground that] requires preparation. What's the difference between giving a lecture and talking about the contents of a lecture? Audiences have notoriously fragile attention spans, therefore it is important that the speaker reach the listener and that the listener understand and retain the message being delivered. This requires a significant amount of preparation, and while a good speaker may sound off the cuff and spontaneous, chances are they have prepared well. The scientific apprentice must consider that each minute at the podium requires at least one hour of thorough preparation. Many speakers confuse entertainment with public speaking; scientists do not need to entertain an audience, but they should captivate them with a world that was previously unknown to them.

Another value of conferences, besides the positive aspect of preparing one and being a lecturer, is attending conferences and actively participating in them. It is important to note here that knowledge is not the only thing that you are seeking at a conference. That is the reason that you can search the Web as much as you like looking for knowledge and information, but human contact is what will provide the final mark. Conferences are a natural environment to socialize and lifelong friendships started in these gatherings can generate research collaborations, job opportunities, and more. As a rule of thumb, the basic interaction of people can be a rewarding one if the researchers are open to the other person and the environment is right. Conferences, even small ones, are sources for meeting other people who might later be your postdoc or your boss or somebody who will make decisions about your future.

9.3. The Optimal Size of the Cancer Research Laboratory

In an analysis of the scientific productivity of nearly 3000 researchers, a relationship between grant size and scientific productivity was found. The resulting plot showed that both measures peaked at around US$750,000 in annual funding; at higher funding levels, the median publication number and average impact factor were both discernibly lower (Wadman, 2011). These data confirm what was already known, that the increase of the laboratory size makes the group harder to manage and as a consequence the productivity falls. Here again the metric is determined by the number of papers or publications. This once more raises the question of the importance of impact of the research outcome. Impact is difficult to measure, e.g., should it be measured by the publications, patents, citations, and employment it garners? And of course the next question is what are the implications of a spike of productivity in a specific area of cancer research? Are the data innovative enough to further the next step of our knowledge? All of these are difficult questions to answer and probably to determine the average laboratory size and funding another metric is needed, a better measure to determine how to distribute the scarce funding sources available to the present generation of cancer researchers. However, one thing that is difficult to deny is the fact that the research funding agencies have long dreamed of favoring scientists who have a proven record of turning their work into tangible benefits for the society and the economy (Gilbert, 2010). Many academics are concerned that the added focus on research impact could skew funding toward applied research and result in decreased interest in our understanding of basic biological concepts.

9.4. How are We Ready to Share?

In the ideal system all our data, scientific reports, essays, memories, comments, and peer and non-peer review publications would be available for everybody to use, enjoy, or criticize. In this ideal world, the government or private research institutions should be the repository for this invaluable human labor. However, in the real world we treasure and do not share those data that we decide are not adequately considered, as by certified publications or peer reviews. The main reason is that we are afraid that this not properly verified data will be used by others, less hardworking laborers who seek quick self-benefit. In other words, we are afraid that our work will be stolen or plagiarized. Probably, those scientists that work with the data generated by others, like biostatisticians or bioinformaticians, who need large databases, will be more open to the idea that all data must be available for scrutiny and further study; however, that idea is not appealing to the biological-oriented researcher who is painfully involved in the data collection. The positive side of a repository is that many observations or data that are available will not be lost; therefore, the crucial part is that the owner of those data is not ignored. In the eventuality that the researchers are eager to share their data, the next problem is what format they should use to storage that information. A significant progress has been the accumulation of gene data sequencing in normal and tumor tissue, e.g., that has created a tremendous advance in our understanding of the role of mutations in cancer. Another example is the Database of Genotypes and Phenotypes (http:// www.ncbi.nlm.nih.gov/books/NBK154410/), which archives and distributes the results of genome-wide association studies, medical DNA sequencing, molecular diagnostic assays, and almost anything else that relates people's traits and behaviors to their genetic makeups. This database allows open access to summaries and other forms of information that have been stripped of personal identifiers. Therefore, these are tangible examples that well-managed databases are important for our advances in the diagnosis, treatment, and prevention of cancer. In a *Science* editorial written by Marcia McNutt and titled "IAmAResearchParasite", the author interestingly refers to the points discussed above and makes a strong encouragement for data sharing (McNutt, 2016). There are millions of unpublished data that could be beneficial to know because the tricky part is that we do not know where the answer is. The building up of this new concept of data sharing needs to be addressed by the present as well as the new cadre of cancer researchers and institutions, both federal and private, must help to develop the infrastructure to accommodate those data, classify them, and also guarantee that the researchers are not lost in the process.

9.5. Creating an Adequate Environment for Academic Cancer Research

There is no doubt that places that have invested in massive infrastructure creating a hub for scientific and technological innovation are the locations that will attract cancer researchers. One important incentive is the creation and availability of fellowships for young cancer researchers and a fostering of translational research that allows these young researchers to see the clinical application of their endeavors. There are a good number of fellowships offered by the NIH, the Department of Defense, as well as other private institutions that are targeted to postdoctoral training and even some of them for undergraduate students. The paradox is that in many cases these positions are not filled simply because the qualifications of the applicants do not meet the requirements or because some of the applicants do not meet the visa status.

It is also necessary to foster more hybrid funds between academy and pharma companies, which allow the training on the know and the how to translate ideas from the basic research laboratories to the industry or drug development, for example. Although there are funds from the government for developing start-up incubators, the difficulty is in the decision to jump from the academic to the industry. This is a jump that is not easy to do and in most cases there is not any net to protect against failure. The presence of academic institutions with adequate ties to industry could provide this safety net for the young research entrepreneur.

Academic institutions must foster the integration of young scientists into the management echelon by integrating them in committees where planning and economic decisions are discussed and studied. The more acquainted the cancer researchers of this generation are with these issues, the better prepared they will be to manage the efforts of small and large laboratories.

References

Gilbert, N. UK science will be judged on impact. *Nature,* **468**: 357, 2010.
McNutt, M. IAmAResearchParasite. *Science,* **351**: 1005, 2016.
Perkel, J.M. The manuscript-editing marketplace. *Nature,* **531**: 127–128, 2016.

Moskovitz, C.C. and Kellogg, D. Inquiry-based writing in the laboratory curse. *Science*, **33**: 919–920, 2011.

Nayar, A. Educating India. *Nature*, **472**: 24–25, 2011.

Pennisi, E. Conquering by copying. *Science*, **328**: 165–167, 2010.

Rumsey, A.S. When we are no more. Bloomsbury Press, USA, 2016.

Russo, J. *The Apprentice of Science: A Handbook for the Budding Biomedical Researchers.* World Scientific, Singapore, 2010.

Wadman, M. Study says middle sized labs do best. *Nature*, **468**: 356–357, 2010.

Chapter 10

What the Future of Cancer Research Will Look Like

10.1. To See the Future We Must Understand the Past

Since 1990, we have witnessed a doubling of funding from the National Institutes of Health (NIH) budget even as the demands for funding grew much faster than the supply. As described in previous chapters, the need for more funding was in a certain way fueled by incentives for institutional expansion, by the rapid growth of the scientific workforce, and by the rising costs of research, as has been discussed extensively in an article by (Alberts *et al.* (2014). As indicated by these authors, the resources available to the NIH are estimated to be at least 25% less in constant dollars than they were in 2003. The consequences of this imbalance include dramatic declines in success rates for NIH grant applicants and diminished time for scientists to think and perform productive work and develop original ideas, less time for thinking, reading, and talking with peers. As Alberts *et al.* indicated:

> *the current system is in perpetual disequilibrium because it will inevitably generate an ever-increasing supply of scientists for a finite set of research resources and employment opportunities. The competitive race created by the low funding from 30% to less than 10% of the grant application has suppressed the creativity, cooperation, risk-taking, and original thinking required to make fundamental discoveries, and biomedical scientists are spending far too much of their time writing and revising grant applications and far too little thinking about science and conducting experiments. This has produced conservative, short-term thinking in applicants, reviewers, and funders. The system now favors those who can guarantee results rather than those with potentially path-breaking ideas that by definition, cannot promise success.*

The consequence of this environment has created a culture where "publishing scientific reports, especially in the most prestigious journals, has become increasingly difficult, as competition increases and reviewers and editors demand more and more from each paper. Long appendixes that contain the bulk of the experimental results have become the norm for many journals and accepted practice for most scientists. This superego need to publish in high impact journals has produced a backlash; there is significant research work that cannot be replicated or confirmed".

This environment of racing for survival has been made even worse by expanding regulatory requirements and government reporting on issues such as animal welfare, radiation safety, and protection for human subjects, to mention just a few of the many administrative burdens that are imposed on researchers. Although some of these requirements are important, these multiple conditions shorten the useful hours scientists have for thinking and working on vital issues like peer review of manuscripts and grant applications, the two most important academic endeavors that a good scientist must perform. Because of these time constraints and the need for the principal investigator to keep the funding running in this hypercompetitive atmosphere, the thinking, study, and peer review of manuscripts and grant applications are in some occasions postponed or delegated to less trained hands. That is easily detected reading the criticisms received from journals or funding agencies.

10.2. Can We Predict the Future of Scientific Research?

On the basis of the exponential technological advances of the last 100 years, we could probably predict a futuristic account of how the science of cancer research will be. However, I am interested in first predicting the future of our humanness. It is not doubted that as a species we have evolved positively through the ages, but since the beginning of written history, the basic nature of us, as humans, has not changed so much. We have been changing our explanation of nature but our basic brainpower has not altered significantly during the last 2000 years of our human history. We still experience the same feelings of hate, love, remorse, and violence that our ancestors had. The best example is in the literature records, from Greek tragedy to Shakespeare on down to the modern writers, our human feelings of love and regret have not changed. Based on this simplistic but accurate reasoning, it leads us to wonder how the technological advances will change us as humans. Will computing power robotics, artificial intelligence, biology, nanotechnology, and three-dimensional printing, to cite a few, change us as humans or only make our lifestyle and survival different? This is the real question that we need to pose to ourselves. In my vision, the technological advances have not only made us better equipped to survive and to live, but also have created a significant disparity between the ones that have and the ones that do not have. Still our basic humanness has not been changed. Therefore, if we stick to this premise the fundamental problem is how the technological advances will help us to cope with the problems of disparity, social unrest, health care, and diseases that are still afflicting millions of people. A more basic question is how we can harness that artificial intelligence and robotic innovation to help us solve and predict catastrophic events.

However, whereas I am not too much concerned about the way that we evolve technologically, in the physical realm of computers, artificial intelligence, and robotics I am concerned about the way that we scientists deal with those technologies that hamper our human fabric. For example, the emergence of a powerful gene-editing technology, known as CRISPR-Cas9, has elicited furious debate about whether and how it might be used to modify the genomes of human embryos. The changes to their genomes would almost certainly be passed down to subsequent generations, breaching an ethical line that up to now was considered a no-trespassing zone. Therefore, we need more than ever to develop the human ethics that set the barriers that we will not cross. If we as scientists create the boundaries that we can cross, it will be significantly more rational than leaving this to nonscientists who will be unable to discern the potential goodness to eradicate some crippling genetic diseases versus the temptation to tamper with our humanness.

10.3. The Future of Cancer Research

According to a report from the Biomedical Research Workforce Working Group, the future of cancer researchers is uncertain due to the slowdown in the employment sectors — academia, government, and the pharmaceutical and biotech industries — that could and should benefit from their long years of training. It is true that the brightest are the ones that at the end are able to occupy a research position in academia or research institutes but they still need to enter the competitive race of funding, and in some cases they will wait 4–5 years before they can get their first award; competition for funding is not an easy game. All of these uncertainties will produce a significant vacuum in future cancer research endeavors. The snowball effect will be measured in years to come.

I will not even dare to speculate which economic and political measurements should be taken to improve the cancer research endeavor; for those who are interested in this aspect, the article of Alberts *et al.* (2014) provides several suggestions that if implemented would be of great help to cancer research in the near and long-term future. What I will emphasize in this chapter are the individual qualities that must be fostered by the future cancer researchers, in part by developing those inner qualities and in part with the help of good mentors in this important discipline of the sciences.

10.3.1. *Look in yourself and find a good mentor*

Before entering the research endeavor, each person must *listen closely to the forces within to discover his or her own special calling,* and to find out as early as possible what they want, and then train to be resilient and focused. These are the two most important qualities one must have to reach their goals and live their dreams. But as important as this advice could be, it is also a mentor's function to recognize the inner calling of future scientists, by seeing beyond appearances to the true scientific mind within. I would like to emphasize especially the point of *learning to see beyond appearances.* This means that the mentor must see the inner person and look past any gender, ethnic, or physical bias. The mentor must meet the mind and soul of the future cancer researcher by listening to them, and only then will the real cancer researcher make him- or herself known. Like a good parent, the mentor should support the budding cancer researcher in any way they can.

As I have already indicated in my book *The Apprentice of Science* (Russo, 2010), each individual is unique to his or her period of time, and there is no particular pattern from which a scientist emerges, at least in the initial stage of their lives. Therefore, we must ask ourselves: How can we recognize budding scientists in their first manifestations? In an era in which science is extremely competitive and requires a significant amount of training, the earlier the individual finds out that science is part of their life endeavor, the better equipped they will be for their personal future and the future of science in their country and around the world.

How do we know when someone has what it takes to embark on the journey of cancer research? There is no fixed set of rules, but the ability to analyze a problem and persist until it is solved is a good beginning. Many major discoveries are made not due to special talents, but rather common sense and hard work. There are individuals with great memories, quick learning abilities, dexterity, and excellent communication skills. Without a doubt these people, if they persevere and focus in their scientific endeavor, can make great contributions. However, most individuals are not endowed with all of these qualities, yet still make great contributions to science by using their talents wisely and pushing themselves to succeed. The key is having curiosity, a genuine desire to know, and not being afraid to ask questions. Curiosity, the ability to think logically, and rigorous work habits are the three main components that will define a cancer researcher's talents. Whether or not a budding cancer researcher has the ability to think logically and independently is easy to detect. Those who consider the written word the final word — in other words, they see no need to reexamine or reinterpret a published paper — probably do not grasp the true definition of "research", which is "a diligent and systematic inquiry or investigation into a subject in order to discover or revise facts". The word derives from a French word, *rechercher,* meaning "to seek". This entails challenging a hypothesis and going even further to consider that there may be other explanations for the same hypothesis. To see a lag in the knowledge, to understand that what is known in a biological process does not cover all angles, and that many questions remain unanswered, is the beginning of the scientific process. The budding cancer researcher must also understand that an inquisitive sense of perception is of great importance for self-discovery, but it is the fitness to persevere with an idea and concentrate on a specific problem that counts in the end (quoted from Russo, 2010). The reality is that an idea or problem changes from beginning to end, and it is the continuous challenge of solving a constantly evolving question that shapes our neural networks. This process requires solitude and on some occasions total isolation from noise and other external distractions. This self-imposed retreat is part of learning intellectual introspection, which is key to the cancer researcher's success.

10.3.2. *Learn how to work in a team*

Preparing one's mind for scientific observation leading to significant discoveries requires three basic things: individuality, teamwork, and the ability to work in solitude. Let's examine these concepts further. What makes a research process unique is the idea generated by one individual (see Chap. 1), meaning one needs to have the drive, constancy, devotion, and persistence to originate a research idea and the vision and foresight to develop it. This requires a significant amount of effort and the stamina to persevere hours, days, weeks, and possibly years in order to pursue this singular goal. It also requires teamwork. One of the things that distinguish modern science from the way it was practiced a century ago is collaborative work. This is mainly because science continues to become more complex, requiring a multidisciplinary approach in which people of different talents, backgrounds, and abilities come together to work on a specific problem. A critical ingredient for successful teamwork is recognizing what each scientist's role is within the team and maintaining an atmosphere of individual thought. Preserving the uniqueness of individual thought in order to apply it to the common goal is what ultimately makes a team successful.

Solitude is the other vital ingredient that makes intellectual individuality and team approaches effective, and in the end, it is a driving force in scientific discovery. Solitude is an environment (both physical and mental) that allows one to contemplate a problem over and over again, to absorb and generate ideas about the end result of a lengthy process. The inability to work in solitude could be a significant detriment to the scientific endeavor. Solitude is related to one's ability to visualize problems and their solutions — not in a singular flash of insight, but on a repeated and continuing basis (see Chap. 8). I would like to emphasize, however, the importance of combining individuality and solitude with teamwork because it is the latter, when one idea is managed from different perspectives, that can help illuminate the final goal of the research process (Russo, 2010).

A good trial of the ability of the new generation of cancer researchers to work as a team will be put to the test in developing the "moonshot project", led by Vice President Biden. This project, formally titled Precision Medicine Initiative, aims to use genomics, informatics, and redesigned clinical trials to revise cancer diagnostics, improve prognostic tools, and develop new therapeutics that will require a significant amount of team of researchers plus the invaluable collaboration of 1 million volunteers, yet to be recruited. If the project is executed in the right way, it could provide a unique set of data that may change the way that oncology is practiced today. Will the Precision Medicine Initiative solve all the problems? I doubt it, but it will provide a new, badly needed infusion of funding in the underfunded research enterprise of today.

10.3.3. *Pursue your original idea*

Finding one's area of interest, or research subject, is one of the most arduous tasks a young cancer researcher will face, yet it is one of the most important in that it will define the nature of the scientist's work for the rest of his or her career. For some scientists, it is a singular idea, hypothesis, or discovery that defines their entire life's work, while others will make multiple contributions to their field (Russo, 2010). It is important that a cancer researcher not be overwhelmed by the achievements of their predecessors but instead keep in mind that even the most revered giants of science at one time struggled with the same uncertainties in finding their intellectual niche. In that sense, the scientists of the 17th century were not so different than those making their way in the 21st century. Although occasionally inspiration for an idea falls serendipitously into a researcher's mind, often ideas begin with suggestions made by a mentor, or the principal investigator of a laboratory. Over time, as the aspiring cancer researcher continues working in a lab, he or she might make an association based on the surrounding research milieu and from there go on to awaken a deeper interest in a given topic. Sometimes it is life experience that draws students to a particular area. Many budding cancer researchers have indicated to me that their desire to study breast cancer stemmed from the loss of a mother, an aunt, or a grandmother to cancer. The emotional and fundamentally instinctual desire to alleviate suffering, as much as the introduction of a challenging idea, could mark the beginning of a lifelong path. Whatever the origin, the birth of a research idea is, in the end, the result of three processes working together: reading, writing, and experimentation. Only when carried out in tandem can the research endeavor and the discovery of oneself as a scientist become possible (Russo, 2010). Cancer researchers must focus their attention on the research idea and not cloud their thinking with peripheral concerns. *Will anyone like it? Is the idea accepted enough by the establishment? Will it win the Nobel Prize?* If there is a genuine desire to research a particular topic, the rest will come on its own.

10.3.4. *The laboratory environment with a look toward the future*

We must give credit to architects for designing the environment of our cities. Over the past four decades, laboratory design has seen improvements. As has been discussed by Goldstein (2006) in a commentary written for Cell, "from obscure quarters, mostly in basements, laboratory design has emerged as a way to foster interaction with other researchers". The emergence of a safety code has also benefited the construction of the modern laboratories by combining technical improvements with flexibility and comfort. The best example of this evolution is in the work of Louis Isadore Kahn, an American architect who created a style that was monumental and monolithic; his heavy buildings for the most part do not hide their weight, their materials, or the way they are assembled. However, the same architect built both the Richards Medical Research Laboratories for the University of Pennsylvania and the research building of the Salk Institute for Biological Studies in La Jolla, California. The former is considered ill-designed because of isolated spaces, dreary lighting, and inefficient organization of space, whereas the latter has generous dimensions, open structural spaces, and flexibility for building modular lab components. The creation of open spaces and the flow of natural light integrated with better maintenance infrastructure have permeated all the new constructions of research laboratories and established a model for future building strategies. The closed spaces are gone and the new lighted and open spaces facilitate better interaction among the researchers and also better sharing of resources (Goldstein, 2006).

As discussed by Goldstein (2006), "modern research labs are designed to simultaneously balance functional needs and safety concerns (always of primary importance) with ergonomics, indoor air quality, thermal/visual/acoustical comfort, flexibility, and the provision of shared social spaces. These interaction zones represent a range of space types such as break rooms, lounges, and other places in which researchers can relax and collaborate outside the lab proper."

Besides these architectural elements I would like to reemphasize my vision, described in my previous book *The Apprentice of Science* (Russo, 2010), that the laboratory is more than a physical workplace, it is not only the room or series of rooms where experiments are planned and conducted, but also the place where science plays out and unfolds knowledge and understanding — just as the researcher and his or her group are defined by more than just what is within the laboratory walls. The laboratory provides the physical tools required for research ideas to flourish, but although they are physically permanent structures, they can be long-lasting ideas developed by the principal investigator. The best example is the breast cancer research laboratory that was created in 1973 and renamed 30 years later when Dr. Irma H. Russo passed away and the Fox Chase Cancer Center designated that laboratory with her name. The Irma H. Russo, MD, Breast Cancer Research Laboratory is a permanent idea that transcends life.

Laboratory and researcher(s) are one, and, as discussed in the previous section, the research idea or theme is the driving force, or the flame, that makes the laboratory a living entity. It is my conviction that a cancer research laboratory must be created with a vision to alleviate cancer suffering and the drivers must maintain the thematic of the lab beyond the architectural needs.

Is there a special recipe for running future cancer research laboratories? I will say yes. The most important component of the research endeavor for the new cadre of cancer researchers is sharing the enthusiasm and desire for the knowledge gained from experimentation with the new generation of scientists. This ability to train and teach the new generation is a major component of what makes a research laboratory successful. Almost all of the world's legendary laboratories have provided the intellectual milieu necessary to foster scientific learning, and in the end, this is one of the research laboratory's essential missions.

10.4. The Future Management of Cancer Research

Cancer research is not only a discipline but also a major force that moves billions of dollars in the United States and around the world. When I joined the American Association for Cancer Research (AACR) in 1973, the number of attendees at their annual meeting was a few hundred researchers and clinicians that in general met in a large hotel in a major American city. In 2016, the AACR gathered more than 15,000 people with more than 5000 abstracts, and the venue is now a major convention center. The AACR is a measure of the importance of cancer research in our society, as is the number of researchers who come to this event from all over the world. Besides recognizing the success of the AACR, the gathering is a metric of dollars needed to keep this research enterprise alive. Several questions emerge from this concern: Is the number of research institutions needed? Is the number of cancer researchers justified? Will there be funding for cancer research available? Do the biomedical enterprises in cancer research have the vitality to break new paths?

While there is not an easy way to answer all these questions, there is some consensus, at least conceptually, that the biomedical research system cannot be expanded indefinitely at a substantial rate (Alberts, 2014). The opinion of some researchers is that the disequilibrium exists because there is an increasing supply of scientists for a limited number of research positions and funding for research. If this is the case, the obvious solution would be to decrease the number of scientists, but it is the opinion of some authors that this is not the root of the problem and that the basic problem is to apply business models of operation to basic science, clinical research, and clinical medicine (Lazebmik, 2015; NIH, 2012).

In trying to find an explanation for this conundrum, a document widely credited for the success of US science over the past 70 years came to my attention (http://www.nsf.gov/abouvhistory/vbush1945.htm). This document was prepared in 1945 for President Franklin Roosevelt by Vannevar Bush, an MIT professor, engineer, and science administrator who supervised most of the US military research during World War II, including the Manhattan Project and the mass production of penicillin. The quote reads:

The Government should provide a reasonable number of undergraduate scholarships and graduate fellowships in order to develop scientific talent in

American youth. The plans should be designed to attract into science only that proportion of youthful talent appropriate to the needs of science in relation to the other needs of the nation for high abilities. In this document Bush's main argument is about the primacy of science, and his vision reached far beyond counting scientists. He emphasized that *scientific progress on a broad front results from the free play of free intellects, working on subjects of their own choice, in the manner dictated by their curiosity for exploration of the unknown...* and noted the complexity of developing scientific talent because *no one can select from the bottom those who will be the leaders at the top because unmeasured and unknown factors enter into scientific, or any, leadership. There are brains and character, strength and health, happiness and spiritual vitality, interest and motivation, and no one knows what else, that must need enter into this supra-mathematical calculus.*

In my perspective, this document is a vital one to keep in mind when developing plans for the future of cancer research. The success will depend on educating the new cadre of cancer researchers with a solid basis of the scientific method and teaching them to think critically and solve problems creatively and collaboratively. As I have described in previous chapters and discussed in *The Apprentice of Science* (Russo, 2010), creative thinking, problem-solving, motivation, and persistence should not only be taught but fostered by emulation incited by their mentors and *again not all our graduates or physicians will be prepared or have the personal motivation to be scientists and that is an important concept that we must hold clear in our minds.* To be a scientist is not a profession, it is special call to unveil the unveiled and pursue the unknown. It is difficult to determine who will make it and who will not, but it is certain that we can see it when it is there.

10.5. To Search for Greatness as the Real Future for Cancer Research

As indicated in the first section of this chapter, we need to understand the path for planning our future to have a sense of aspiration; the notion of coming from little knowledge and making something meaningful is an important driving force for a cancer researcher. The other lesson that we must learn is that there are very smart people from whom we can learn and that the leadership they emanate is very personal and unique. To know these people not as remote icons but as real actors in the drama of cancer research is inspiring. It is always difficult to define greatness, but those people who develop strength, resilience, and intuition through the years are the ones most likely to show greatness and help us understand what it is. The new cadre of cancer researchers needs to learn how to find and breathe this sense of greatness. Those that have it are brilliant thinkers with a clear and thought-through research approach. They also establish rapport with different people and are able to easily communicate with those whose knowledge is limited as well as those who are their peers.

Greatness is not associated with brashness, loudness, or heavy-handedness, and great cancer researchers are not afraid of the critics. There is a sense of nobility in their appearance and behavior. This nobility is not about being nicer or better than anyone else, rather it is a feeling about the research endeavor that transcends gender, race, or social conditions.

References

Alberts, B., Kirschner, M.W., Tilghman, S., and Varmus, H. Rescuing US biomedical research from its systemic flaws. *Proc. Natl. Acad. Sci. USA.* **111**: 5773–5777, 2014.

Goldstein, R.N. Architectural design and the collaborative research environment. *Cell,* **127**: 243–246, 2006.

Lazebmik, Y. Are scientists a workforce, or how Dr. Frankenstein made biomedical research sick. *EMBO Rep.* **16**: 1592–1600, 2015.

National Institutes of Health (NIH). *Biomedical Research Workforce Working Group Draft Report.* Bethesda, MD: National Institutes of Health, 2012.

Russo, J. *The Apprentice of Science: A Handbook for the Budding Biomedical Researchers.* World Scientific, Singapore, 2010.

Index